COLOR THEORY AND MODELING FOR COMPUTER GRAPHICS, VISUALIZATION, AND MULTIMEDIA APPLICATIONS

THE KLUWER INTERNATIONAL SERIES IN ENGINEERING AND COMPUTER SCIENCE

COLOR THEORY AND MODELING FOR COMPUTER GRAPHICS, VISUALIZATION, AND MULTIMEDIA APPLICATIONS

by

Haim Levkowitz
University of Massachusetts Lowell
Lowell, Massachusetts, USA

KLUWER ACADEMIC PUBLISHERS
Boston / Dordrecht / London

Distributors for North America:
Kluwer Academic Publishers
101 Philip Drive
Assinippi Park
Norwell, Massachusetts 02061 USA

Distributors for all other countries:
Kluwer Academic Publishers Group
Distribution Centre
Post Office Box 322
3300 AH Dordrecht, THE NETHERLANDS

Library of Congress Cataloging-in-Publication Data

A C.I.P. Catalogue record for this book is available
from the Library of Congress.

The publisher offers discounts on this book when ordered in bulk quantities. For more information contact: Sales Department, Kluwer Academic Publishers, 101 Philip Drive, Assinippi Park, Norwell, MA 02061

Printed on acid-free paper.

Printed in the United States of America

CONTENTS

LIST OF FIGURES

Chapter 8

Chapter 9

Chapter 10

Chapter 11

LIST OF TABLES

PREFACE

Imagine a world in which the colors you see do not necessarily represent the correct ones. Imagine, when you approach a traffic light, you cannot tell what color the light is. Imagine the colors of traffic lights changing from one block to the next one. Could you?

If you can't, just go to your computer. In the world of visual computing, you cannot be sure what the colors will be. Even today. And today, the awareness of computer professionals to color has grown orders of magnitude.

When I got introduced to some interesting color problems in medical imaging back in 1984 (and thus reintroduced to color, which I had researched and explored in the seventies), it seemed like nobody really bothered; there were very few publications that attempted to understand the relationship between color and the computer.

Within a few years, that number grew significantly. Finally, people were beginning to realize that, just as colors play such an important role in our day-to-day life, they are important on our computer displays. And, thus, color should be studied, and the appropriate usage on computer displays should be explored, understood, and *practiced.* I cannot say we have reached that stage, yet, but I am glad to report that progress has been made.

This book is the result of over a dozen years of research, teaching, consulting, and advising, first at the Medical Image Processing Group, Department of Radiology, the University of Pennsylvania, and then here at the Institute for Visualization and Perception Research and the Graphics Research Laboratory at the University of Massachusetts Lowell.

I started putting the material presented here together the first time for a half-day color tutorial I taught at the IEEE Visualization '91 conference in San Diego. The success of that tutorial suggested that it should be expanded from a half-day to a full-day course, and that inviting other color experts to contribute their expertise would strengthen it. As a result, I invited Philip Robertson (then of CSIRO, Canberra, Australia, and now with Canon Research in Sydney) and Bernice Rogowitz (IBM Research, Yorktown Heights, NY) to join me. We took our "show" to SIGGRAPH '92 in Chicago, and again to IEEE Visualization '92 in Boston. Phil's and Bernice's contributions to the tutorial, and thereby to this book, were numerous and significant. I am grateful to them for lending me their vast knowledge and experience.

Since then I have taught a similar tutorial at several other conferences, and have used parts of this material in my visualization tutorials, as well as in my computer graphics and visualization courses here at Lowell.

Over the years, many people have asked me for advice about their color graphics applications. At some point, I got convinced that many others, who haven't had the opportunity to take the tutorial, and who couldn't ask me questions, could use some of the material presented here. Thus came the idea for of this book.

The goal of this book is not to teach you everything you need to know about color; numerous publications are available to do that. It is not even to teach you everything you need to know about color computer graphics; no, there aren't numerous publication available to do that. The goal of this book is to convince you that it is important to pay attention to color (if you are not convinced yet); to give you the basic fundamentals of color vision in general, and as related to visual computing in particular; to make you aware of the repercussions of your color choices; and to stimulate you to further explore this topics. One not-so-hidden agenda: I'd like to encourage students to pick color as their doctoral research topic.

To help us accomplish these goals, we have established a World-Wide Web (`http://www.cs.uml.edu/~haim/ColorCenter`) site, where you can find all the color images, more information, examples, tools, and links to other related sites. As all good Web sites should be (forever), this site is under construction, and will hopefully remain so. What you will find on the Web site are all the color figures in the book, additional example images, recommended color scales, some useful software tools, an extended, annotated bibliography, and links to other relevant sites.

We would like you to visit the site, and take advantage of what is available there. To make the most of it, we suggest the use of an automatic monitor (such as URL-minder at `http://www.netmind.com/URL-minder/URL-minder.html`) to alert you to changes in the site.

We would also appreciate your comments, suggestions, contributions, and critique.

This book would not have materialized if not for many people who have helped me, directly or indirectly, make it happen.

More than fifteen years ago, my colleague and friend Yair Censor influenced to a large degree the direction of my academic career. Gabor Herman, my doctoral dissertation advisor, colleague, and friend, made significant contributions to the material presented in this book. Ron Pickett found me at a job I did not enjoy, and together with Georges Grinstein and Stu Smith, brought me over to Umass Lowell, where I have had a wonderful time ever since. Giam Pecelli, our previous Department Chair, and Jim Canning, our current Chair, have both provided me enormous support here; they, together with the rest of the Computer Science Department faculty, have made it so wonderful to work here.

Many extremely capable and dedicated students have worked over the years on projects, the results of which can be found in this book and on the Web site: Kerry Shetline wrote the first versions of MCMTRANS and MCMIDS (and suffered through my notorious bug-finding ability). Nupur Kapoor and Bogdan Pytlik implemented the first version of the color icon. David Gonthier developed the new version of the color icon, and incorporated it in NewExvis, our visualization environment. He was helped by Jude Fatyol, Omar Hoda, and Lisa Masterman. Rob Erbacher developed the parallel implementation of the color icon and the GLHS color model. Krishnan Seetharaman added three-dimensional spaceball navigation to the GLHS color model, and made numerous other contributions, too numerous to list them all. Alex Gee, Pat Hoffman, J.P. Lee, Dave Pinkney, and Marjan Trutschl, have all helped make life at IVPR as pleasant as it is, in many different ways, but mostly in their eagerness to do whatever is necessary, whenever it is necessary. When one is blessed with having so many wonderful students, it is inevitable to forget someone; my apologies to those I have not mentioned: your contributions have been as important and as appreciated.

My family in Israel, Colombia, and here in the US, all deserve my thanks for all they have given me over the years.

Finally, no thanks could be enough to Ea, Merav, and Shir, with whom I should be spending this weekend, instead of in front of my computer, writing this. It's wonderful to have you!

Lowell, Massachusetts *Haim Levkowitz*
January 1997

PART I

COLOR THEORY

1

HUMAN VISION

The rays, to speak properly, are not coloured. In them there is nothing else than a certain Power and Disposition to stir up a Sensation of this or that Colour. (Newton [New04], pp. 124–125.)

What is color? Is it a property of the object that we see? Is it a property of our visual system? Of light?

It is all of the above. Color is our perception, our response to the combination of light, object, and observer. Remove either one, and there is no perception of color. That means that in order to fully use color, we have to understand all of these.

And, we have to make sure that visual technologies that we develop are matched to the human visual capabilities.

1.1 THE HUMAN INTERFACE

Print technology has evolved over centuries. That gave us sufficient time to understand how to best present information in print so that the reader will get the most out of it, with the least amount of effort.

It is no coincidence that most books have similar formats; the type of fonts used, their size, the number of words per line, the number of lines per page, have all been optimized for the reader. It took those centuries to learn this. But because the technology was not developing at a very rapid pace, we had those centuries to learn.

On the other hand, the electronic display revolution has happened orders of magnitude more rapidly. It only took 30 years from the first utilization of CRT displays till today's virtual reality head-mounted displays!

This has not provided sufficient amount of time to fully assess the technology, and understand what the optimal ways to present information are.

In addition, the technology has not reached the point where we can deliver those parts we know are necessary. For example, today's best displays—CRT and flat panel—do not possess sufficient spatial resolution to provide the same image quality that photography offers. In the color domain, similar constraints are still limiting color resolution and fidelity. Moreover, as we will see later, different color display devices provide rather incompatible color gamuts.[1] Thus, you are not guaranteed that a color image will appear the same (or even similar) on two displays. Color modeling that will provide cross-device color compatibility is still a topic of research [Xu96].

1.1.1 Requirements for image quality

The short list of requirements for adequate image quality includes:

inxximage,quality

Sufficient luminance and contrast. Both good luminance and contrast are absolutely essential for a good quality image. Poor luminance or contrast will render an image hardly usable, or even completely useless.

No flicker. The primer current display devices utilize *refreshed* technology, i.e., the contents have to be updated (or "refreshed") continuously to maintain the image. This causes periodic changes of the overall luminance level of the image to fluctuate between bright and dark. The frequency of this fluctuation determines whether our visual system will perceive *flicker* (the periodic bright-dark change) or will not. Flicker is perceived at frequencies under 50 cycles per second, but also depends on other conditions of the display device, as well as the observer.

Minimized effects of spatial sampling. All display devices sample continuous images and convert them to discretized ones. The size of the

[1] The *gamut* of a color display device is the set of all colors that the device is capable of displaying.

smallest picture element, *pixel*, of a discretized image determines the effects of the sampling (also known as *aliasing*). Aliasing, its effect on images, and anti-aliasing techniques have been discussed in the computer graphics literature, see, e.g., [FvFH90].

Perceptually lossless image compression. As the use of images, and their size, increase, compression becomes more and more important. To achieve high compression rates, it is usually necessary to use lossy compression algorithms. The loss of information by such algorithms may or may not be perceived by a viewer. It is essential that whatever loss is incurred during the compression process will not affect critical decision making, such as is often required in medical visualization, to name but one application.

Convincing impression of depth. Two-dimensional images are often used to convey three-dimensional situations. This is done by creating an impression (an "illusion") of depth.

Effective use of color. Being our main topic, we have left this as the last item.

1.1.2 Technology questions hinge on visual perception

In order to be able to provide the correct technological answers to accomplish image quality as described above, we need to understand how the human observer perceives visual information. More specifically, we must be able to answer the following questions:

- How do we process luminance, contrast, color, and motion?
- How do these mechanisms constrain our choice of how to capture, sample, compress, and display information?

1.1.3 Some definitions

We now define a few terms we will be using.

Physical stimulus: measurable properties of the physical world, such as luminance, sound pressure, wavelength.

Sensation: the immediate effect of the physical stimulus.

Perception: the effect of sensory phenomena that are also mediated by higher-level processes, such as memory, attention, and experience.

Psychophysics: the study of the sensations and perceptions that physical energies—such as brightness, loudness, and color—produce.

Neurophysiology: the study of the physiological mechanisms mediating the transduction, coding, and communication of sensory information, [Hub88].

1.2 COLOR AS A TRI-STIMULUS MEDIUM

Color is a sensation produced in the brain in response to the incidence of light on the retina of the eye. The sensation of color is caused by differing qualities of the light emitted by light sources or reflected by objects. It may be defined in terms of the *observer*, in which case the definition is referred to as perceptual and subjective, i.e., it depends on the observer's judgment. Or, it may be defined in terms of the *characteristics of light* by which the individual is made aware of objects or light sources.

However, the definition in terms of the characteristics of light is related to the sensation experienced by the observer. That is, the light received by the retina is composed of a spectrum of different energies in different wavelengths. Only at the eye, and further at the brain, the particular spectrum is translated to the experience of a particular color. Moreover, many different spectra are perceived as the same color. This observation—referred to as *metamerism*—was made by Newton in his 1704 book *Opticks* [New04, von70].

It is commonly accepted that both specifications can be accurately defined in a three-dimensional space, that is, by specifying *three components.*

1.2.1 Perceived color: Specified in terms of the observer

The three components that specify color in terms of the observer are:

Hue: Specifies the actual "color" that we see (red, yellow, etc.).

Intensity/lightness/brightness/value: Specifies the achromatic (luminance) component, which is the amount of light emitted or reflected by the color. It can be thought of as "how much black is mixed in the color." The term *intensity* refers to achromatic colors.[2] The term *lightness* refers to objects, and is associated with reflected light. Verbal degrees for lightness are {*very light, light, medium, dark, very dark*}.[3] The term *brightness* is used for light sources, and is associated with emitted light. It can take any of the degrees {*very dim, dim, medium, bright, very bright*}. The term *value* was first used by Munsell in the *Munsell Book of Color* system [Mun05, Mun69, Mun76]. It refers to the relative darkness or lightness of the color, in the Munsell system.

Color scientists commonly use *lightness* and *brightness* interchangeably, but much less so with *intensity.*[4]

Saturation/chroma: Specifies purity in terms of mixture with white, or vividness of hue. It is the degree of difference from a gray of the same lightness or brightness. *Saturation* is colorfulness relative to the color's lightness while *chroma* is colorfulness compared to white. When lightness changes a change in saturation is perceived. Increased lightness causes a perceived decreased saturation, and vice versa. Chroma, which is associated more with the Munsell system, does not change with lightness. Verbal degrees of saturation are {*grayish, moderate, strong, vivid*} [BBK82, W. 80].

1.2.2 Color specified in terms of light

The three components that specify color in terms of light are:

Dominant wavelength: Specifies the actual "color" that we see, and corresponds to the subjective notion of *hue.*

Luminance: Specifies the amount of light or reflection. For an achromatic light it is the light's *intensity.* For a chromatic color it corresponds to the subjective notion of *lightness* or *brightness.*

---

[2] Although intensity is a physical quantity that has measurable physical energy units (cd/m^2), it is sometimes used interchangeably with the other terms, which are perceptual quantities, and have psychophysical units.

[3] These can be quantified using a rating scale.

[4] We use the term *lightness.* We have chosen it rather than *brightness* (which is somewhat more appropriate for our main focus—color monitors) to avoid notational ambiguity between B for blue and B for brightness.

Physical	Perceptual (psychophysical)
Luminance	brightness (light sources; emitted light)
	lightness (objects; reflected light)
Dominant wavelength	hue (hue circle; red vs. purple, etc.)
Purity	saturation (vivid vs. pastel)

Table 1.1 Corresponding physical (light-based) and perceptual (observer-based) terms for color description.

Purity: Specifies the spectral distribution that produces a certain color of light. It is the proportion of pure light of the dominant wavelength and white light needed to define the color. Purity corresponds to the perceptual notion of *saturation.*

A particular color can be described uniquely by three components even though it is a response to light, which is an infinite-dimensional vector of energy levels at different wavelengths. This means that the mapping between the spectral distribution of light and the perceived color is a many-to-one mapping. Thus, many spectra are perceived as the same color. (This phenomenon is referred to by color scientists as metamerism.)

1.2.3 The relationship between observer-based and light-based specifications

The relationship between the subjective and the objective specifications can be described as follows. Among all the energy distributions that cause a particular color sensation, there is one distribution that can be described in terms of the three components of light characteristics above. For each particular objective triple (dominant-wavelength, luminance, purity), there exists a subjective triple (hue, lightness, saturation) that gives the closest description to the sensation caused by the class of distributions represented by the objective triple, see Table 1.1 and, e.g., [FvFH90].

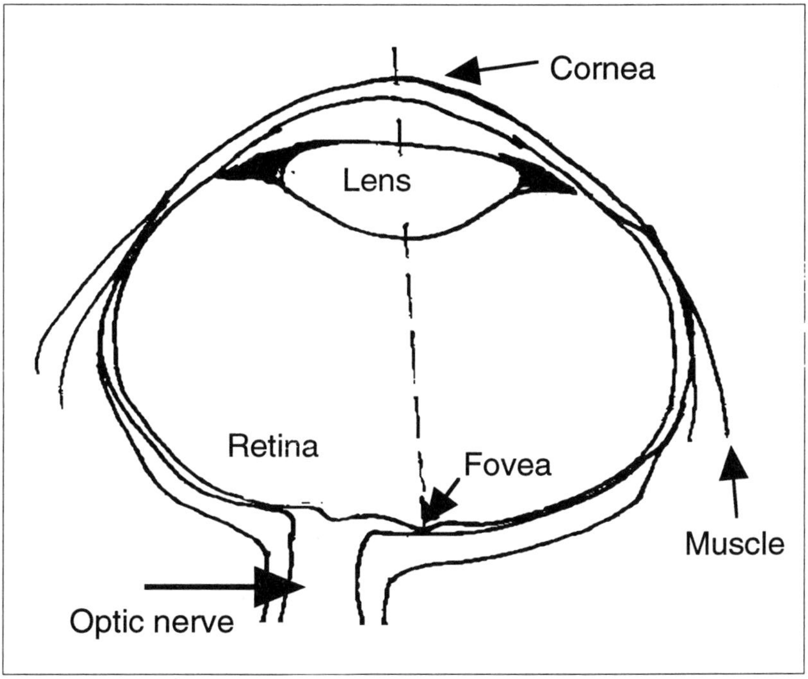

Figure 1.1 Schematic of the eye.

1.3 A TOUR OF THE HUMAN VISUAL SYSTEM

1.3.1 The eye

The front-end interface of our visual system is the eye, see Figure 1.1. The eye consists of several components. The pupil controls the amount of light admitted to the eye (a camera's aperture is modeled after the pupil). Two lenses, the cornea, which is fixed, and a variable-focus lens, provide distance adaptation. The retina, located at the back of the eye, provides the first layer of "image processing" of our visual system.

The retina

The retina contains five layers of cells, in charge of several early image processing tasks.

The first layer contains four types of photoreceptors. These are light sensitive cells, grouped to filter different light phenomena. Approximately 120 million *rods*, which are achromatic light sensitive cells (i.e., they only see "black and white"), are responsible for night and other low light level vision. Day time color vision is provided by approximately 8 million *cones* of three types, which operate like filters for different ranges of wavelength:

- S-type cones: short wavelength, with peak sensitivity at 440 nm (violet, but often referred to, erroneously, as "blue")
- M-type cones: medium wavelength, with peak sensitivity at 550 nm (often referred to as "green" but perceived as yellowish-green)
- L-type cones : long wavelength, with peak sensitivity at 570 nm (yellow, often referred to as "red")

Cones are mainly concentrated in the central vision center of the retina, in particular in the *fovea* (pit—the central one degree of vision); rods are mainly concentrated in the periphery of the retina. There are no photoreceptors (causing a "blind spot") in the optic disk, where the optical nerve connects photoreceptors in the retina to ganglion cells in the brain, transferring image signals to the brain for further processing.

Four other classes of cells in the retina handle image compression and lateral inhibition. Since these are beyond the scope of our discussion, we do not provide any further details. The interested reader is referred to the vision literature (e.g., [Hub88, Mar82, Sch90a, SB90, Wan95]).

Eye movements

It has been proven that eye movement is essential for human visual perception; without constant eye movement, the external world fades away from the viewer.

Six muscles point the eye to areas of interest, at a rate of four cycles, *saccades*, per second. *Minisaccades* keep the eye in constant movement to maintain a clear stable impression of the external world at all times, see for more details [Sch90a, SB90].

1.3.2 Sensitivity vs. resolution

The human visual system provides the capability to adapt for trade-offs between sensitivity (the ability to extract information out of low levels of luminance) and resolution (the ability to distinguish—resolve—small spatial detail).

A one-to-one mapping of photoreceptors (cones) in the fovea to ganglion cells provides the highest spatial resolution (*acuity*). The result is the ability to resolve small spatial detail, but only at sufficient luminance levels.

A many-to-one mapping of photoreceptors (rods) in the periphery of the retina provides the highest luminance sensitivity. The result is a much stronger sensitivity to light-dark changes in our peripheral vision, at the expense of the ability to resolve small detail. The periphery of the retina also exhibits a greater temporal sensitivity, i.e., better sensitivity to luminance changes over time. See the cited vision literature for more details.

1.3.3 The neural pathways from eye to brain

Once the image has been captured in the eye and filtered in the retina, several pathways from the eye to the brain provide support for various tasks.

The eye-superior colliculus pathway controls eye movements and directs our gaze to movements in the periphery. The eye-lateral geniculate pathway includes two pathways: The Magnocellular pathway handles motion information, while the Parvocellular pathway takes care of color and high spatial resolution information.

In the brain, the Striate cortex (in the back of the head, in the occipital lobe) contains the first binocular cells, which are driven by inputs from the two eyes. This is the first place where information from the two eyes is combined to recognize higher-level objects. Objects detected at this level are lines, bars, and blobs. This area also provides spatial-frequency and orientation tuning.

Visual information is delivered throughout the brain via parallel paths; 60% of the brain receives visual input, which is the most dominant input to decision making, memory, and concept formation processes in human beings.

Therefore, looking at—and thus, generating—a display involves more than simply understanding questions of image quality and detectability. It involves understanding how humans seek out, understand, and use information.

1.4 BASIC VISUAL MECHANISMS

The human visual system comprises many different mechanisms. The earliest, and most basic ones involve luminance and contrast perception.

1.4.1 Early vision: luminance perception

Luminance perception is accomplished by a sensitivity to a broad range of luminance variations. The human visual system is sensitive to a range of 14 log-units (i.e., 14 levels in a logarithmic scale, each level ten times higher than the previous one), from the luminance of a dim star, to that of the bright sunlight. However, at any given moment, the sensitivity is limited to a window of two log-units, matched to the ambient illumination. Any luminance levels below the lower level of the window are perceived as the darkest perceptible luminance level, while any levels above the upper level are perceived as the brightest. This provides the dynamics of light-dark adaptation. Typical to all psychometric functions, the apparent brightness is not a linear function of the luminance, rather, it is a logarithmic-like relationship, see Figure 1.2.

Psychometric functions show acceleration, then diminishing returns. In a logarithmic range, equal steps in perceived brightness require geometric increases in values of luminance. This can be thought of as a "perceptual gamma function." As we will see later, early luminance non-linearity is incorporated in many color metrics.

1.4.2 Contrast and spatial resolution

Next to luminance perception is contrast perception. One definition of contrast, C, the ratio between the object's luminance $\ell(O)$ and the background's luminance $\ell(B)$.

$$C = \frac{\ell(O)}{\ell(B)}$$

Contrast sensitivity depends on the spatial distribution of light and dark regions [Sch56]. The minimum modulation required for detecting a grating patterns is a tuned function of the spatial frequency, the *contrast sensitivity function (CSF)*, with its peak sensitivity at two-to-four cycles per degree, and with decreased sensitivity for lower (broader) and higher (finer) spatial frequency patterns, see Figure 1.3.

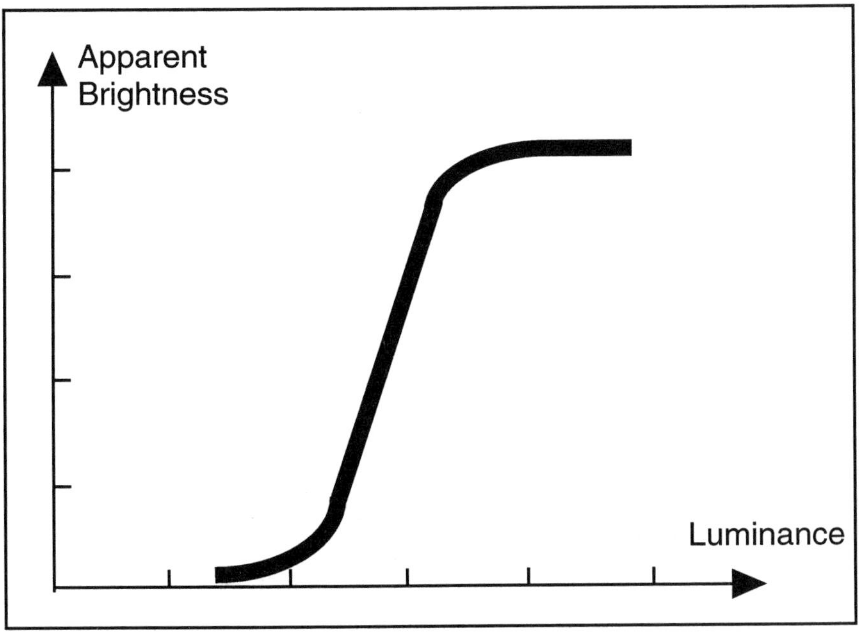

Figure 1.2 Luminance response function.

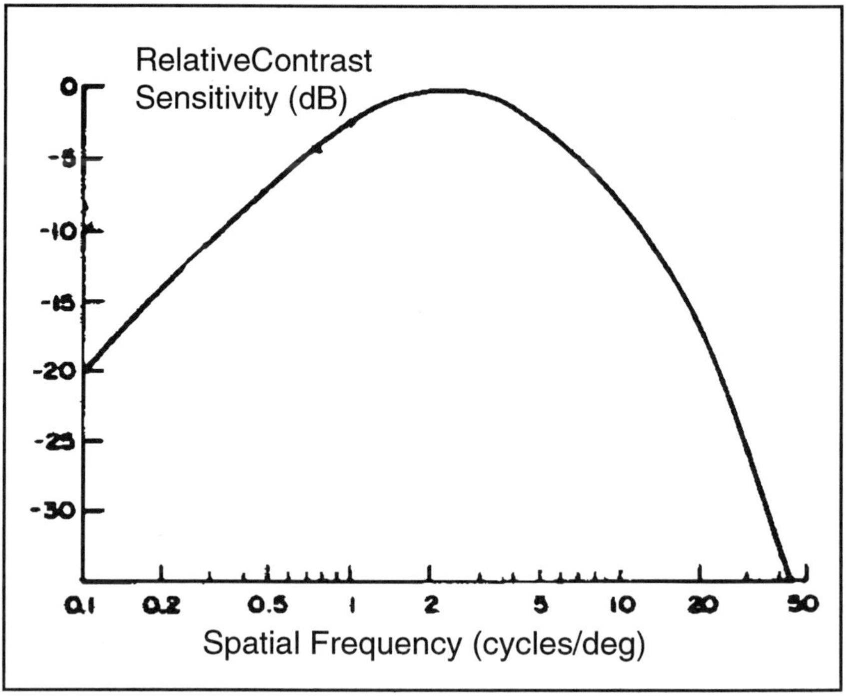

Figure 1.3 Contrast sensitivity function.

1.4.3 Image applications of the contrast sensitivity function

Besides being interesting in and of itself, the contrast sensitivity function provides help with a number of imaging applications. In image coding, efficiency can be gained by devoting the greatest bandwidth of data to regions of the greatest spatial-frequency sensitivity. In digital halftoning, one can hide sampling noise (dotted patterns) using regions of the lowest contrast sensitivity. Measuring display quality, one can evaluate the display modulation transfer function (MTF) with the human contrast sensitivity function (CSF).

However, in utilizing the contrast sensitivity function, one should be aware of two caveats:

1. The shape of the CSF depends on luminance, color, and temporal modulation.
2. The CSF is really the envelope of a set of underlying narrow-band spatial-frequency sensitive mechanisms.

1.4.4 Flicker

One direct result of humans' luminance and contrast sensitivity is the perception of flicker (e.g., in early films and in CRT displays). Perceived flicker depends on luminance factors, not on color. Factors affecting perceived flicker include display parameters, such as luminance, refresh rate, interlaced vs. non-interlaced scanning, phosphor persistence, and display size; viewing conditions, such as peripheral vs. foveal gaze; and observer factors, such as age, caffeine consumption, depressants.

Rogowitz and Park [RP89] conducted a psychophysical experiment, the results of which reveal the relationship between display parameters and perceived flicker. The perceived flicker equation (CM) for interlaced displays is given by:

$$CM = 1.75FR + 15.35P - 31.37 \log L_B - 24.53C - 59.08.$$

For non-interlaced displays it is given by:

$$CM = 2.49FR + 15.35P - 31.37 \log L_B - 24.53C - 59.08,$$

where FR is the field rate, P is the display phosphor persistence, L_B is the background luminance, and C is a constant, $C = 1$ for a color display, $C = 0$ for a monochrome display.

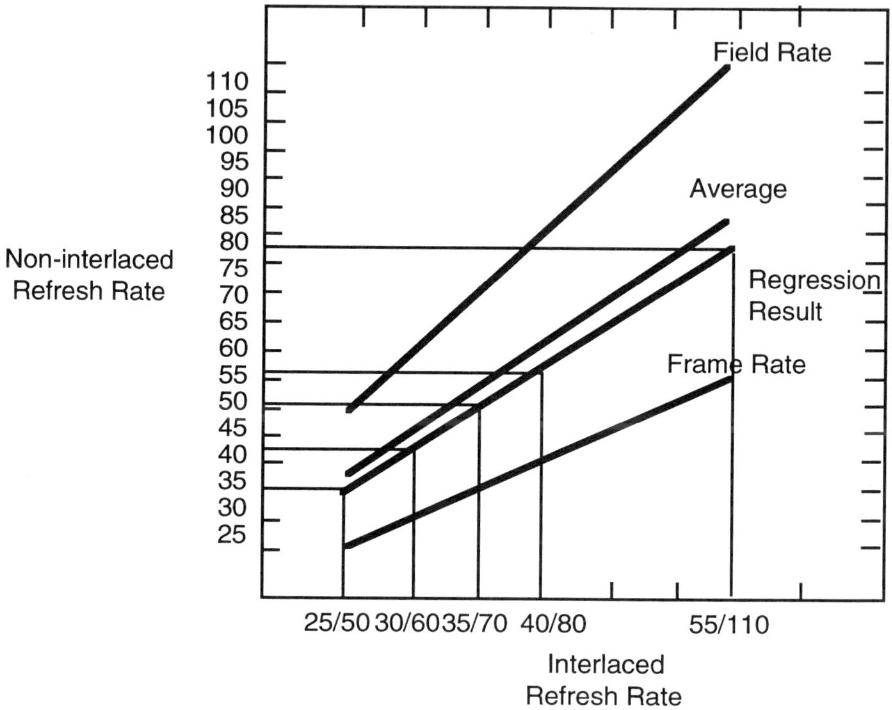

Figure 1.4 Interlaced vs. non-interlaced flicker (from Rogowitz and Park, 1989).

These relationships are valuable for evaluating the benefit of various design changes, and for comparison of flicker in interlaced vs. non-interlaced displays, see Figure 1.4.

An inexpensive way to reduce perceived flicker

One result of this research is this inexpensive method to reduce perceived flicker. Adding a dark glass "anti-glare" faceplate reduces the luminance and increases the contrast. The ambient light is attenuated twice, whereas the emitted light is attenuated only once, thereby reducing the perceived flicker.

1.5 INTRODUCTION TO HUMAN COLOR VISION

Now that we have covered the basic aspects of luminance and contrast perception, we are ready to study human color vision.

1.5.1 Color vision: trichromacy

As we have already mentioned previously, it is accepted among color scientists as well as psychologists and others who have studied human color vision, that color is a tri-stimulus phenomenon, i.e., human color perception is a three-dimensional space. Color scientists often refer to this as *trichromacy.*

Trichromacy starts at the retina, where the cones provide three broadband filters, tuned to three overlapping ranges of wavelength, often referred to as the "blue," "green," and "red" filters, though these are misnomers, as we will demonstrate shortly.

First, the color names do not represent the precise colors where these filters peak, respectively. The so-called "blue" mechanism peaks at a wavelength we would call "blue," but the "green" and "red" mechanisms are largely overlapping, and peak at two close wavelengths we would probably call "yellow" or "greenish yellow."

More important, each cone filter mechanism by itself is "color blind," see Figures 1.5-1.7. For example, in Figure 1.5, the signal out of a particular cone filter would be indentical whether it is the result of the component on the left of the peak of the filter, or the one on the right.

The essence of trichromacy is that the perceived hue depends on the three-dimensional vector of signals detected by the three cone mechanisms in combination.

Due to the many-to-one mapping from the infinite-dimensional light vector to the three-dimensional vector at the retina, the sensation of any color can be created by exposing the three cones to many different three-dimensional vectors (metamerism). This means that even with trichromacy, there are ambiguities, but with fewer cone mechanisms, color perception becomes deficient or dissappears altogether (that is the essence of color deficiencies).

For example, equal excitations of the medium and long wavelength mechanisms by a bichromatic light at 530 nm and 630 nm (see Figure 1.6) will produce

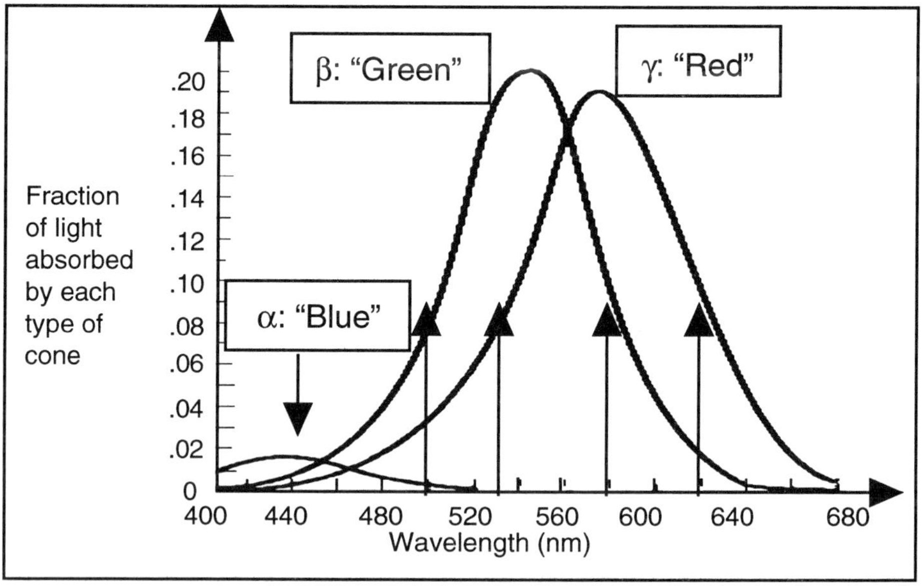

Figure 1.5 Individual cone mechanisms are color blind.

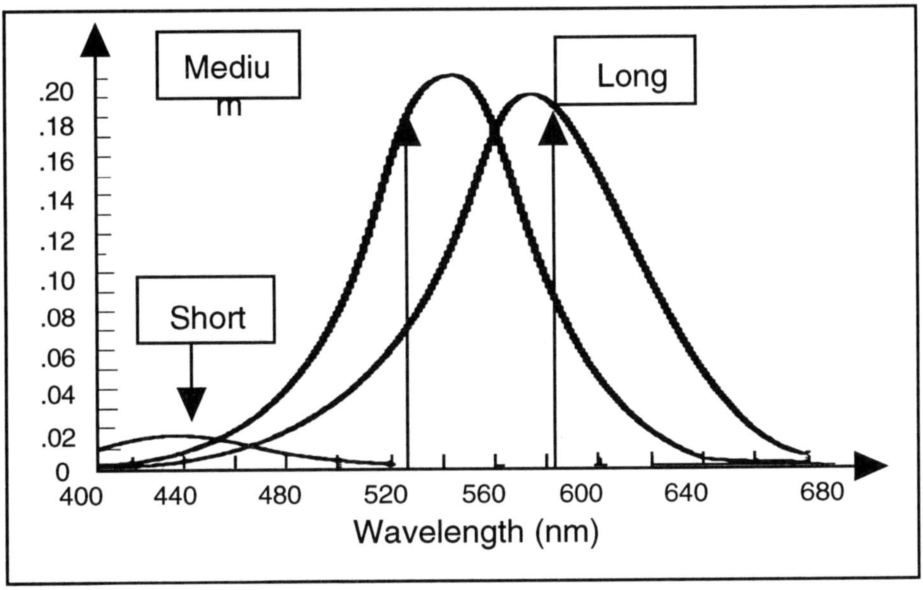

Figure 1.6 A bichromatic yellow: equal light components at 530 nm and 630 nm.

the sensation of "yellow," the exact same sensation that will result from an excitation by a monochromatic light at 550 nm, (see Figure 1.7).

1.5.2 Implications of trichromacy

The important results of trichromacy are that any hue can be matched by the combination of three primaries, and that any hue can be produced by an infinite number of wavelength combinations. This is the basis for color television and color video display terminals (VDT) technology in general.

In order to produce millions of colors one needs only three primaries. For example, the sensation "yellow" does not require a separate "yellow" gun in

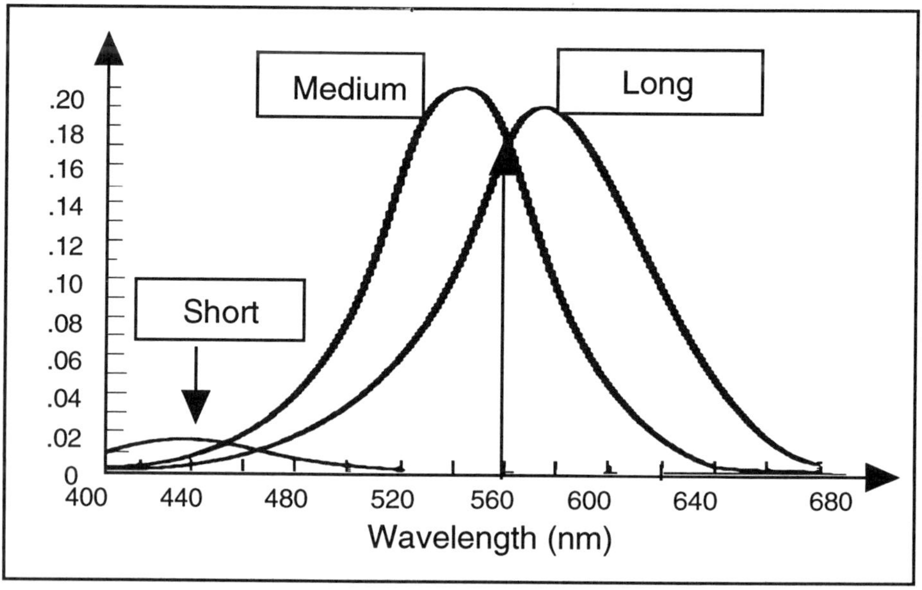

Figure 1.7 A monochromatic yellow: light at 550 nm.

the color monitor; it can be produced from the appropriate proportions of the three guns ("red," "green," and "blue"), which makes it a very efficient process.

For any color system, the *primaries* of the system define the range, or *gamut*, of colors the system can produce.

Primaries and color mixing

Any three colors that are linearly independent (i.e., cannot be mixed from one another) can be primaries.

Additive mixing is the process of mixing the emissions of light sources that cover different parts of the spectrum, where black is the result of no colors mixed in (zero energy), white is the result of mixing of maximum amounts of the three primaries (maximum energy). Color television is an example of additive mixing. The Red, Green, and Blue (RGB) are most commonly used as additive primaries.

Subtractive mixing is the process of filtering the reflection of parts of the spectrum. Here white is the result of no mixing (the entire spectrum is reflected) while mixing the maximum amounts of the three primaries yields no reflection of light at all, i.e., black. Color printing is an example of subtractive mixing. The Cyan, Magenta, and Yellow (CMY) are the most commonly used subtractive primaries.

Another implication of trichromacy is that it is relatively easy to perform color difference measurements and calculate color differences (typically referred to as ΔE).

1.5.3 Second stage: opponent processes

The photoreceptor outputs go through a recombination process in the optic nerve, where they are converted into three *opponent channels* (Figure 1.8):

1. $R + G$: The achromatic contents of the color (lightness/brightness). Note that blue is excluded from this channel; indeed blue does not contribute to the perception of lightness. Later we will see that as a consequence of this, changes in blue only are not sufficient to convey perceived changes in color, and thus are not appropriated for coding data variations.

2. $R - G$: One of the two chromatic contents channels of the color (referred to as "red-or-green" or "red-minus-green"). This channel represents the fact that humans do not experience color relationships that can be described as "reddish-green" or "greenish-red" (compare to "yellowish-green," "greenish-yellow," "greenish-blue," "bluish-green," "reddish-blue," "bluish-red," which can all be experienced).
3. $Y - B$: The other chromatic contents channel (referred to as "yellow-or-blue" or "yellow-minus-blue"). This channel represents the fact that humans do not experience color relationships that can be described as "yellowish-blue" or "bluish-yellow" (compare again to the list above).

1.6 COLOR DEFICIENCIES

Approximately 8% of the male population (less for non-Caucasians than for Caucasians), and slightly less than 1% of the female population suffer from some genetic color deficiency.[5]

The most common deficiency (5% of males, 0.5% of females) is *deuteranomaly*. Deuteranomaly is an anomalous trichromacy, caused by an abnormal M-type cone, resulting in abnormal matches and poor discrimination between colors in the medium (M) and long (L) range of wavelengths. The typical result of deuteranomaly is what is referred to as red-green deficiency: the inability, or at least a great difficulty, to discriminate reds and greens. The abnormal M-type cone has its peak sensitivity much closer to the peak of the L-type cone, thus the two overlap more than in normals, causing significantly reduced discrimination.

A similar deficiency can be caused by an abnormal L-type cone. In this case the peak of the L-type cone is shifted closer to the M-type cone, causing again reduced red-green discrimination.

A more severe case that causes red-green deficiency is *deuteronomy*, or a complete lack of the M-type cone. Similar deficiency can be caused by a complete lack of the L-type cone, see Figures 1.9 and 1.10.

These, as well as other deficiencies, which are caused by a missing or an abnormal particular cone type are much less common [SB90].

[5] These are the people we colloquially refer to as "color blind," a rather incorrect label. Truly color blind people have no color perception whatsoever (they see the world in shades of gray), are very rare, and usually have this deficiency as a result of a head trauma they have suffered.

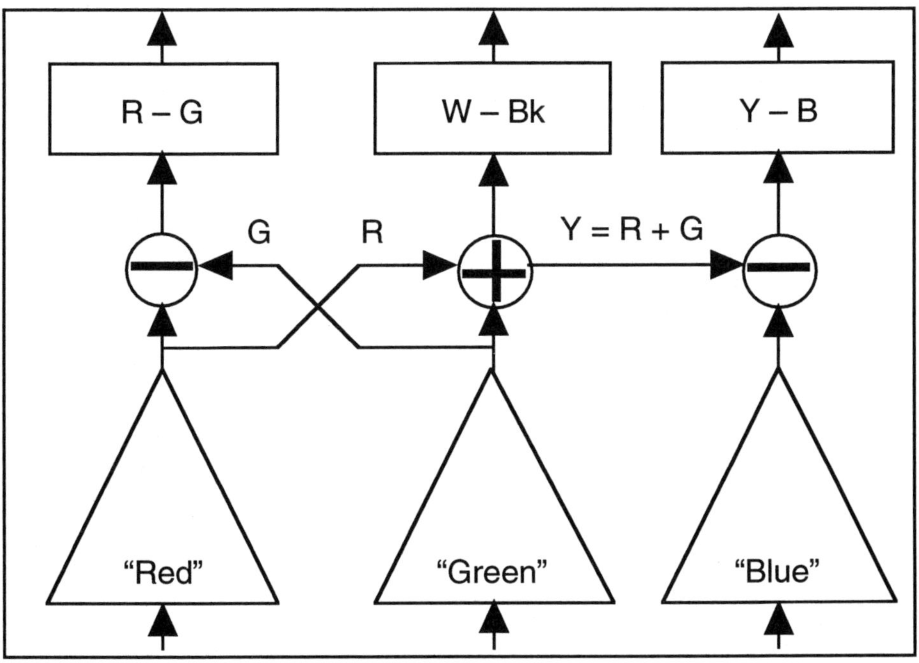

Figure 1.8 Photoreceptor output recombination into opponent channels.

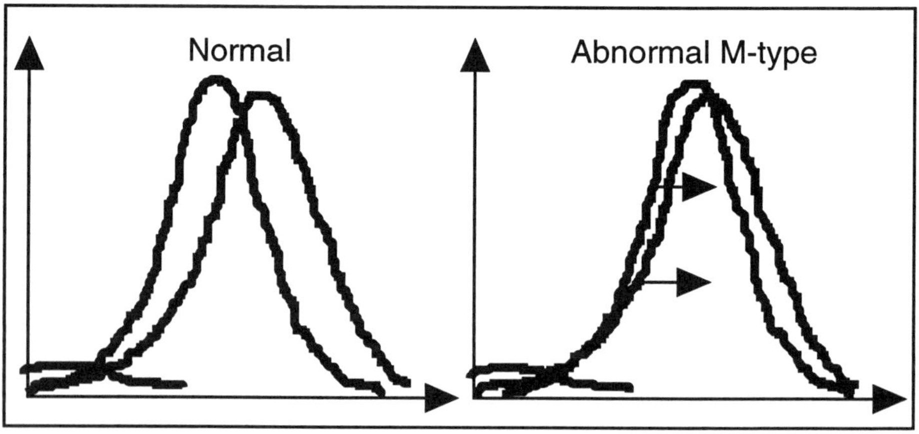

Figure 1.9 Abnormal M-type cones, compared to normal.

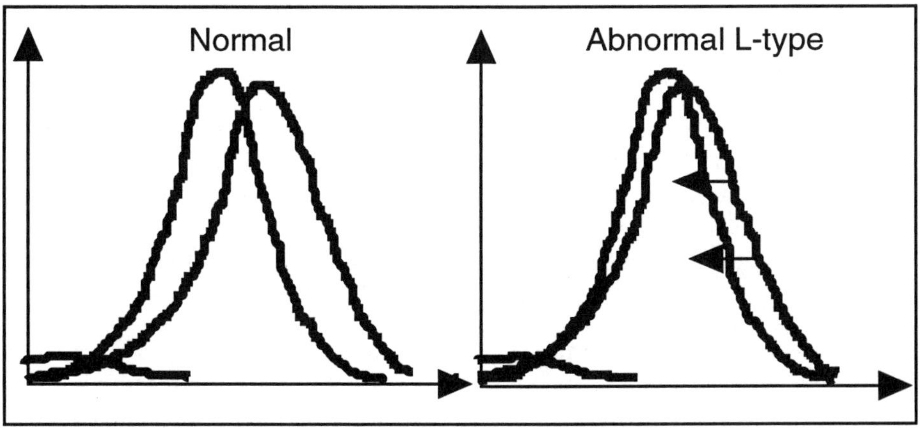

Figure 1.10 Abnormal L-type cones, compared to normal.

1.6.1 Color deficiencies and visual displays

Understanding the nature (and prevalence) of color deficiencies is helpful in designing displays that are useful to as many users as possible, including (at least the majority of) color deficient ones.

A simple rule-of-thumb for creating displays that are usable by color deficient viewers is to always code important distinctions in the image with a redundant luminance cue. For example, a World-Wide Web search tool should highlight found words on the browser in both different color and different brightness.

In addition to redundant coding using luminance, it is strongly recommended not to code differences using colors along either one of the main opponent-processes chromatic channels, in particular the red-green channel, since red-green is the most common color deficiency. In other words, for example, instead of coding ranges of values as shades of green and red, or shades of yellow and blue, select an axis that is a combination of the red-green and the yellow-blue channels, and then code the information along that axis [MG88].

Color deficiency is defined in terms of color discrimination tasks. A color-deficient person will perceive certain colors as being identical, which a color-normal person will be able to distinguish as different. With decreased red-green sensitivity (the most common), colors that are primarily defined by their red or green components, such as rose, beige, and moss green, may appear identical, since the color-deficient observer is less sensitive to the red and green that distinguish them. And many color deficient people are not even aware of their deficiency.

However, color naming may not be impaired in color-deficient observers since it depends on learning. Also luminance, and other cues can contribute to the ability of color-deficient observer to call colors by their correct names (for example, red is dark, yellow is bright). It is well known that red-green deficient drivers use the physical position of the red light at the top and the green light at the bottom of traffic lights to distinguish them. The (rare) practice of placing traffic lights horizontally remove this additional physical clue.

1.7 COLOR-LUMINANCE INTERACTIONS

Color and luminance perception are two complementary mechanisms in the human visual systems. They complement each other in their capabilities, mostly with respect to resolution.

Luminance vs. color resolution

The luminance system can resolve very fine spatial variations. Its sensitivity peaks at two-to-four cycles per degree, with a cut-off frequency of 60 cycles per degree.

The color system can resolve only course spatial variations. Its peak sensitivity for iso-luminant gratings is at the low end of the spatial frequency spectrum, with a cut-off frequency between 10 and 20 cycles per degree. For the differential spatial resolution of color and luminance systems, see Figure 1.11.

To put it in different words, we can make three basic observations:

1. High spatial frequency sensitivity is mediated by the luminance mechanism.
2. The luminance mechanism has a greater bandwidth.
3. Low spatial frequency sensitivity is mediated by the color mechanism.

Let us examine these in turn.

The luminance mechanism mediates high spatial-frequency tasks

High spatial-frequency contents (that is, small samples, such as text and line drawing) are hard to discriminate in the lack of sufficient luminance contrast. For example, it is difficult to detect yellow text on a white background as there is little luminance difference (that is, low contrast) between the yellow text, which is a high spatial-frequency component, and the white background.

High spatial resolution depends on luminance, and is independent of hue.

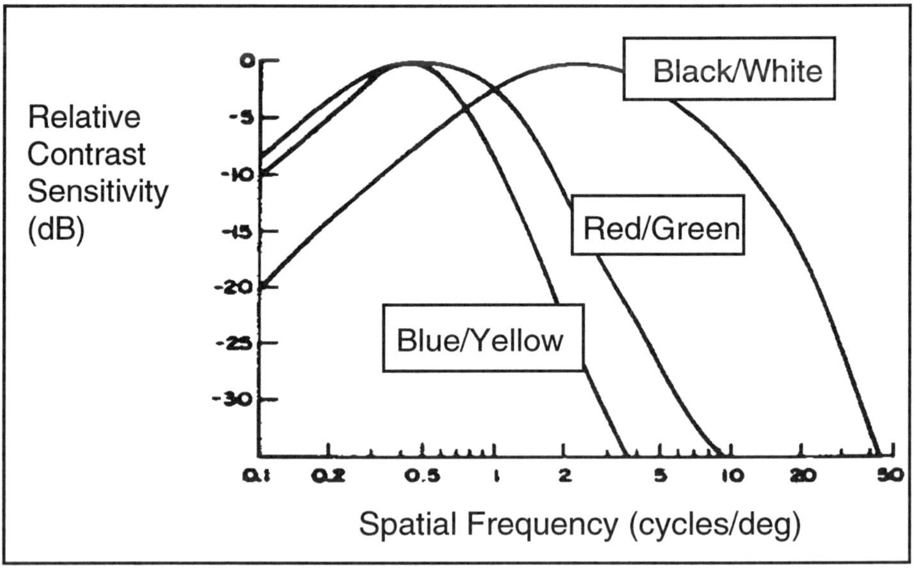

Figure 1.11 Modulation required to detect chromatic and achromatic grating patterns as a function of spatial frequency.

The luminance mechanism has a broader bandwidth

More bandwidth is required to encode spatial variations of luminance. Thus, adding color to convert black-and-white television into color television required little additional bandwidth—the color information was just squeezed in-between the bands. An important lesson is that image compression schemes devoting most bandwidth to luminance achieve higher compression rates while maintaining higher image quality.

The color system is more sensitive to low spatial frequencies

At the extreme, small color targets will "lose" their color, and will look achromatic, see Figure 1.12. On the other hand, colors look more saturated and intense over large areas. For example, the same color that looked quite subdued on a small paint chip at the paint store, will look very intense on a large wall; intensely colorful natural scenes (such as sunsets) can look wan in snapshots, but impressive on a large screen.

This means that pictures on the large screen of a High Definition Television will appear more colorful, thus part of its appeal.

It also means that while designing graphical user interfaces (GUI), small hue differences between windows in the interface are sufficient to distinguish them; there is no need for jarring, saturated colors.

1.8 SUMMARY AND NOTES

We have covered quite briefly the important foundations of human vision in general, and human color vision in particular. We have tried to emphasize, where applicable, the implications of various properties of the human visual system on the tasks that are at the focus of this book, namely, computer-based image generation, manipulation, and presentation. This is by no means an exhaustive description of the human visual system; such description would occupy (and have occupied) many volumes. For detailed descriptions of the human visual system, the reader is referred to the general vision and color literature, including (but not limited to) [JW75, Sch90a, SB90, Wan95, WS67].

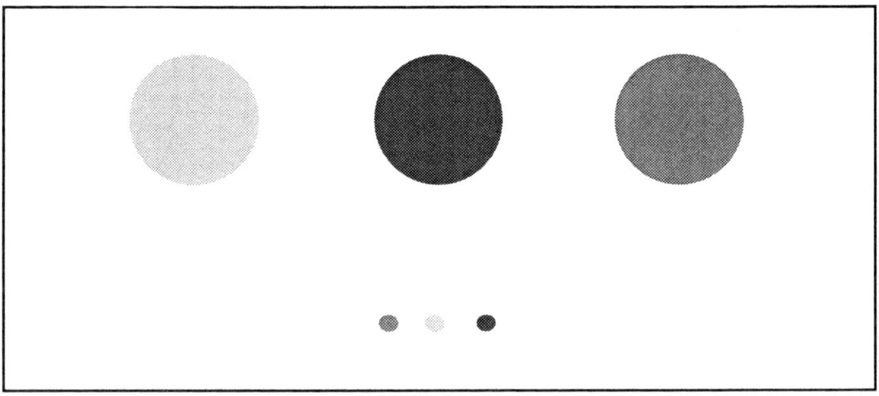

Figure 1.12 Size and Perceived Color. At a distance the color of the small circles is hard to discriminate. (See Web site for color version.)

2

COLOR ORGANIZATION AND COLOR MODELS

In this chapter we discuss color organization and modeling, which is the attempt to provide a way to describe the set and order of colors perceived by an observer, or those a particular device can produce.

2.1 INTRODUCTION TO COLOR MODELING

A *color model* (also *color solid*, *color space*) is a three-dimensional body used to represent some color organization according to a particular choice of three coordinates that describe color. The attempts to organize colors in some order can be traced back to Leonardo da Vinci's *Notebooks* around 1500. Since then, many have tried to organize colors in different solid shapes. Early models varied in shape from pyramids to cones to spheres, as well as some irregular shapes. Historical details can be found in [Ago79, Hes84, Ost69, Rob84, Wri84]. Most of the models in use today (e.g., the *Munsell* color system [Mun05, Mun69, Mun76]; the *Ostwald* color system [Ost31, Ost69]; the more recent *Natural Color System (NCS)* [HS81] and the *Colorid* system [Nem80]; the Optical Society of America (OSA) system and several models used in computer graphics) are all based on similar concepts, and have color solids that can be continuously deformed into the color sphere proposed by Runge in 1810 in his book *Die Farbenkugel* (The Color Sphere) [Run73].

The basic concept, which is mutual to all these models, is continuous variations along the three axes of the model (such as hue, saturation, lightness), see Figure 2.1. Combined with upper and lower bounds on the values along these axes, these yield a three-dimensional color solid.

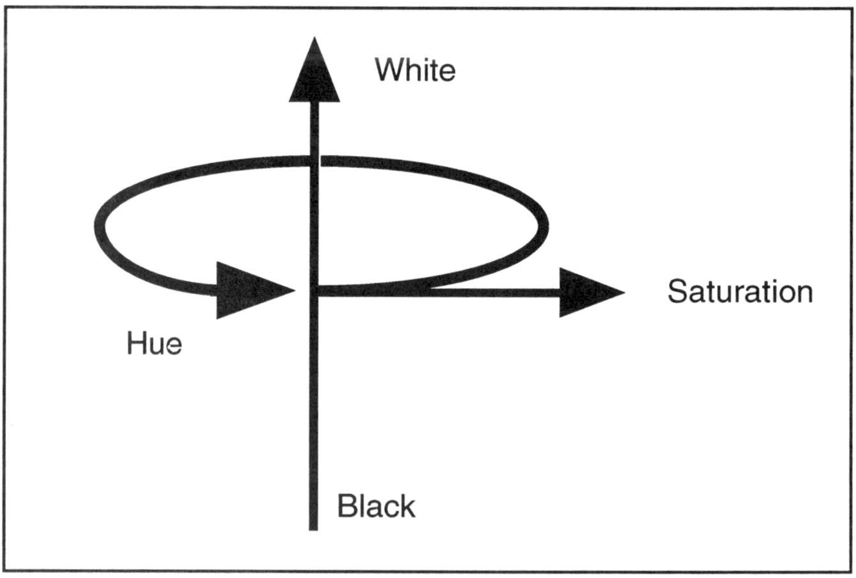

Figure 2.1 Continuous variations in hue, saturation, lightness.

A number of considerations are necessary in order to maximize the perceptual functionality of a color model. Since most human beings, most of the time think about colors in terms of hue, lightness, and saturation (in this order) it is important to assess any model with respect to the following considerations:

- Sufficient separation of hues (hues are typically specified in angular measures).
- Sufficient relative separation in saturation and lightness.
- Useful coordinate systems, both perceptually (human terms) and computationally (machine terms).

We discuss these with respect to the various specification system we describe here.

2.2 OVERVIEW OF COLOR SPECIFICATION SYSTEMS

The color models mentioned above are part of a broader hierarchy of color specification systems, which have been in use in a wide range of applications. Table 2.1 summarizes the different categories of color specification systems. In the following, we discuss some of these in more or less detail. Due to our focus on color in computer graphics related applications, we dedicate the majority of Chapter 3 to those systems that have direct applications in computer graphics and related disciplines.

Our goal is to order colors in a rational system while providing as close a model as possible to the way humans perceive colors. We examine several color ordering systems.

2.3 PROCESS-DEPENDENT SYSTEMS (INSTRUMENTAL)

These are systems that model the gamuts (or color sets) of specific instruments, such as color monitors and printers.

Specification system	Example
Physical system Instrumental system	RGB (CMY)
↓	↓
Colorimetric system	CIE XYZ
↓	↓
Perceptual order system (Uniform)	Munsell, CIELUV, CIELAB Hue, Lightness, Saturation
↓	↓
Natural (naming) system	red, yellow, green, blue light, mid, dark

Table 2.1 Color specification systems summary.

We divide them into two groups, additive, which operate on the basis of light addition, and subtractive, which operate on the basis of light subtraction (or filtering out).

Additive systems

Additive systems, such as stage lights and color television monitors, generate colors by additive mixing of light from (typically) three light sources, such as individual lights or electron guns.

Such systems are typically modeled by the RGB color cube, described in detail in Chapter 3. However, these models suffer from a number of shortcomings:

1. Instruments specifications vary greatly. For example, different color monitors may have different phosphors. Thus, a model that accurately describes one instrument, is very unlikely to provide an adequate description of another one, even within the same category (such as color monitors), let alone across types.

2. From a perceptual point of view, the relationships between the instrument's coordinate system and a perceptual one that humans are comfortable with (usually of the hue, lightness, saturation type) is not readily available.

3. Distances within the model do not indicate the size of color differences, thus the model does not provide accurate color difference metrics.
4. These devices are not capable of producing the entire gamut of colors perceptible to humans. There is no clear treatment (usually none at all) of those colors that a particular device cannot produce.

Subtractive systems

Subtractive systems—such as most hard copy devices—operate essentially the opposite way of additive systems. The default situation is one that reflects the entire spectrum back, a white page (compare to the default additive situation, where no energy has been added, and thus the initial color is black). By applying pigments, which are filters that prevent certain parts of the energy spectrum from being reflected back, certain colors are "subtracted" from the default white, thus causing the perception of those colors that have been preserved.

These instruments are most commonly modeled by the CMY color cube, which is the exact same cube as the RGB cube, but using the cyan, magenta, and yellow corners of that cube as *primaries* instead of the red, green, and blue. (We provide a full mathematical description of these models in Chapter 3.)

Thus, the CMY model suffers from the same problems as the RGB model. In addition, however, imperfect dyes and filters may cause crosstalk (which sometimes demonstrate itself as "bleeding" of colors). And, aesthetical as well as economic reasons cause the need for a separate black channel (referred to as "K," thus naming it the CMYK model). The colors these devices produce also depend heavily on illumination.

A typical problem in both additive and subtractive systems is the lack of an easy answer to the question "what are the coordinates of a particular color, say, a mid-orange?"

2.4 PROCESS-ORDER SYSTEMS (PSEUDO-PERCEPTUAL)

Process-order systems (also called *pseudo-perceptual*) partially alleviate the problem mentioned last in the previous section.

These systems use axes that may represent intuitive concepts—such as hue, lightness, saturation (HLS)—but they are not based on psychophysical realizations of HLS. Thus, they are only *pseudo-perceptual*, see Figures 2.2, 2.3, and 2.4. We discuss some of these in greater detail in Chapter 3.

2.5 COORDINATE SYSTEMS BASED ON HUMAN VISUAL MODELS

Opponent-process models

Many models are based on opponent-process formulations (see Section 1.5.3), where the achromatic component $(R + G)$ represents lightness or brightness, and the "ratio" of the chromatic components $(R - G,\ Y - B)$ defines hue. The combination of the chromatic and achromatic components define saturation [29].

One difficulty in such systems is that they do not embody experimentally-observed perceptual attributes of color discrimination as a primary consideration.

CIE Coordinates Systems

The CIE coordinates systems have been accepted by color scientists as the standard for objective, device-independent color specifications.

These coordinates were derived through color matching experiments, which yielded color matching functions to describe an "average" observer with normal color vision, utilizing the CIE Tristimulus Values X, Y, and Z.

The initial result was what is known to be the CIE 1931 Standard Observer: $x(\lambda), y(\lambda), z(\lambda); (2^o)$. Additional development yielded the CIE 1964 Standard Supplementary Observer: $x_{10}(\lambda), y_{10}(\lambda), z_{10}(\lambda); (10^o)$. The former describe "average" normal color vision for a field of view of 2°; the latter is extended to a field of view of 10°.

Together they provide a reliable, reproducible, precise specification of normal color vision. They have been defined to have a direct relationship with the luminous efficiency function V_λ.

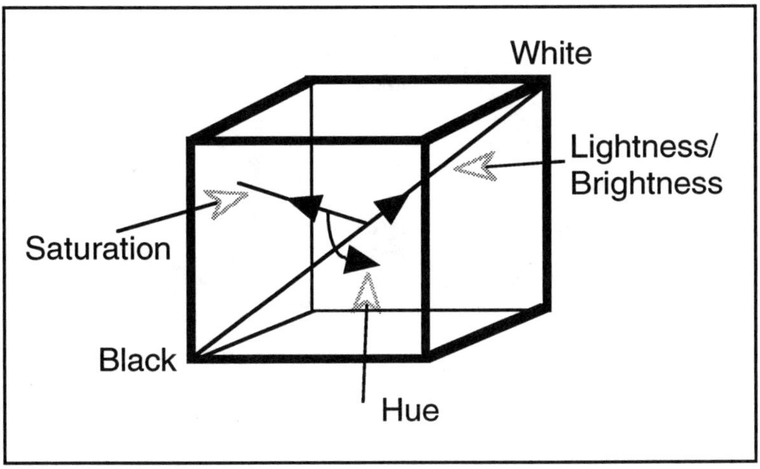

Figure 2.2 Pseudo-perceptual order systems: HLS in RGB cube.

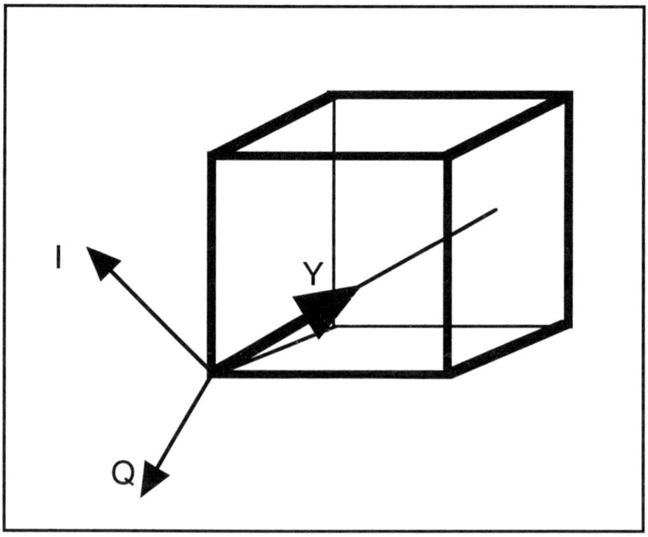

Figure 2.3 Pseudo-perceptual order systems: YIQ in RGB cube.

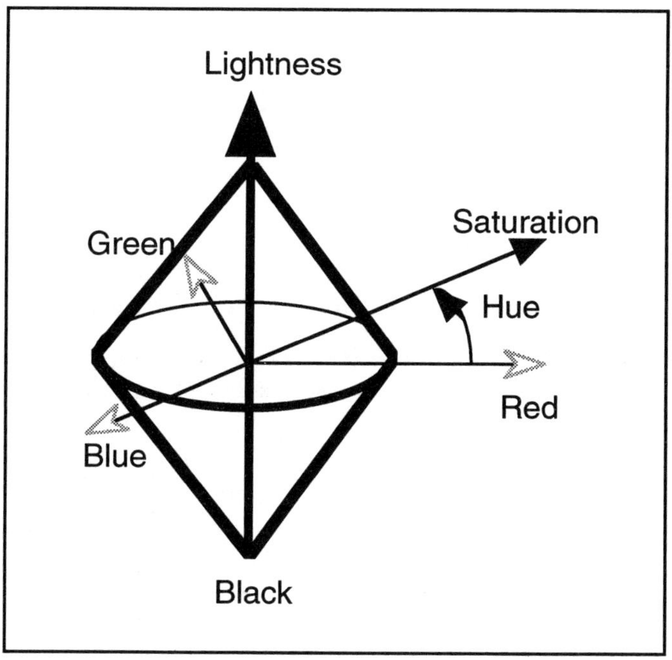

Figure 2.4 Pseudo-perceptual order systems: HLS in double cone.

But, the tristimulus values

$$X = \int_\lambda S(\lambda)x(\lambda)d\lambda,$$

$$Y = \int_\lambda S(\lambda)y(\lambda)d\lambda,$$

and

$$Z = \int_\lambda S(\lambda)z(\lambda)d\lambda,$$

do not provide a useful indication of *color difference* sizes, or intuitive perceptual interpretation (that is, in terms of HLS). The chromaticity coordinates x, y provide only a rough separation into achromatic/chromatic.

Note also that these specifications do not incorporate surround/adaptation conditions, or other perceptual effects.

2.6 PERCEPTUALLY UNIFORM SYSTEMS

Perceptually uniform systems have two basic characteristics:

1. They provide perceptual (HLS) ordering and thus addressing.
2. They provide uniformity: distance in the coordinate system indicates the size of perceived color differences uniformly over the whole color "space."

Two approaches have been taken to develop perceptually uniform systems:

1. Build a color order system experimentally and use it as a reference (for example, the Munsell Book of Color, see Section 2.7, [Mun05, Mun69, Mun76]).
2. Develop an analytical formulation based on discrimination experiments (such as, CIELAB, CIELUV, see Sections 2.7, [CIE78, Taj83a, Taj83b]).

2.7 UNIFORM COLOR SPACES (UCS)

Uniformity is the property of a color space where equal metric steps in the space correspond to equal perceptual steps.

The Munsell Book of Color

The *Munsell Book of Color* is an empirical organization of colors based on human perception. It was derived empirically by Munsell—based on visual judgment—to be uniform. Colors are organized in a way that appears to represent uniformity in human color perception more accurately [Fel89, JB89, Mun76]. The space is composed of color chips organized in equal perceptual steps along its three numerically labeled axes HVC: hue, value (lightness), and chroma (saturation). Any color can be specified by a unique HVC value combination. When the three axes span three dimensions in space, the resulting solid is a distorted Color Sphere (*Die Farbenkugel*, [Run73]). The total of forty Hue values are arranged in a circle; the original Munsell Book of Color was divided into ten sectors (red, yellow-red, yellow, green-yellow, green, blue-green, blue, purple-blue, purple, and red-purple), each further subdivided into ten sectors, a total of 100 equal parts. Values are arranged vertically from 0 (black) to 10 (white). Chroma values are arranged in a radial direction horizontally from the achromatic (Value) axis. The number of chroma steps varies for different hues and values. The Munsell book is made of physical color chips organized in hue pages. For each hue, a page shows the colors of the various values and chromas that are perceivable for that hue. The Munsell book of color is periodically updated. The current total number of chips is approximately 1600 [Mun05, Mun69, Mun76]. Transformations exist to convert color coordinates between H, V, C and CIE x, y, Y.

The CIELUV Uniform Color Space

The CIELUV Uniform Color Space was developed by the CIE, and was recommended for modeling additive light source stimuli. It is claimed that perceived differences between colors are well represented by the Euclidean (square norm) distance in these coordinates [CIE78, Taj83a, Taj83b].

Each color $c = (r, g, b)$ in RGB space has a unique representation in CIELUV, $(L^*(c), u^*(c), v^*(c))$. The LUV coordinates are calculated as follows:

$$L^*(c) = \begin{cases} 25(100 \cdot Y/Y_0)^{1/3} - 16, & \text{if } Y/Y_0 > 0.008856, \\ 903.29(Y/Y_0), & \text{otherwise,} \end{cases}$$

$$u^*(c) = 13L^*(u' - u'_0),$$
$$v^*(c) = 13L^*(v' - v'_0), \tag{2.1}$$

where

$$u' = \frac{4X}{X + 15Y + 3Z},$$
$$v' = \frac{9Y}{X + 15Y + 3Z}, \tag{2.2}$$
$$X = 0.62r + 0.17g + 0.18b,$$
$$Y = 0.3r + 0.59g + 0.11b,$$
$$Z = 0.066g + 1.02b, \tag{2.3}$$

and Y_0, u'_0, v'_0 are the values for the reference white illuminant. An example for a reference white illuminant is the so-called *illuminant-C*, defined as a black-body radiator at $6504^o K$, and for which $Y_0 = M$, $u'_0 = 0.201$, and $v'_0 = 0.461$ [JW75, Taj83a, Taj83b].

The CIELAB Uniform Color Space

Similar to CIELUV, the CIELAB Uniform Color Space was developed by the CIE. But it is recommended for modeling reflected light conditions.

Each color $\mathbf{c} = (r, g, b)$ in RGB space has also a unique representation in CIELAB, $(L^*(\mathbf{c}), a^*(\mathbf{c}), b^*(\mathbf{c}))$. Like in CIELUV, it is claimed that perceived differences between colors are well represented by the Euclidean (square norm) distance in these coordinates [CIE78, Taj83a, Taj83b].

The LAB coordinates $(L^*(\mathbf{c}), a^*(\mathbf{c}), b^*(\mathbf{c}))$ are calculated as follows:

$$L^* = \begin{cases} 25(100 \cdot Y/Y_0)^{1/3} - 16, & \text{if } Y/Y_0 > 0.008856, \\ 903.29(Y/Y_0), & \text{otherwise,} \end{cases}$$

$$a^* = 500[(X/X_0)^{1/3} - (Y/Y_0)^{1/3}],$$
$$b^* = 200[(Y/Y_0)^{1/3} - (Z/Z_0)^{1/3}],$$

Y_0, u'_0, v'_0 are the values for the reference white illuminant, see [CIE78, Taj83a, Taj83b].

CIELUV	CIELAB
$H = \arctan(v^*/u^*)$	$H = \arctan(a^*/b^*)$
$S = \sqrt{u^{*2} + v^{*2}}$	$S = \sqrt{a*^2 + b*^2}$

Table 2.2 Hue, lightness, and saturation in CIELUV and CIELAB.

Hue, Saturation, and Lightness in CIELUV, CIELAB

While CIELUV and CIELAB are claimed to provide an accurate representation of color, as perceived by humans, they do not provide a very intuitive one. It is not trivial to find common color locations as it is not immediately clear what the axes actually represent. A careful observation of these uniform spaces shows that they are organized after opponent-processes models: L^* is the lightness (achromatic) axis, while u^*/a^* $(R-G)$ and v^*/b^* $(Y-B)$ are the chromatic opponent-processes axes, where positive u^*/a^* values represent reds, negative ones represent greens, positive v^*/b^* values represent yellows, and negative ones represent blues. But this is not immediately obvious, and has never been explicitly mentioned in these spaces' specifications. Even after observing the opponent-processes nature of CIELUV and CIELAB, it is still useful to obtain a relationship between these spaces' coordinates and the most common perceptual axes, lightness, hue, and saturation. See Table 2.2 for these relationships.

Other analytically-defined uniform color spaces

Euclidean distance uniformity in CIELAB (and, similarly, CIELUV) suggests that, for example

$$d_{LAB}((50,0,99),(50,0,100)) = d_{LAB}((50,99,0),(50,100,0)) = 1.$$

But, experimental results show that

$$d_{LAB}((50,0,99),(50,0,100)) \approx 5 \times d_{LAB}((50,99,0),(50,100,0)).$$

Luo and Rigg [LR86] proposed L_s, a_s, b_s, to improve the uniformity of L^*, a^*, b^*. Their proposed $CMC(l:c)$ space exhibits good distance (difference) results, but the relationships are non-Euclidean; L_s, a_s, b_s achieves similar distance results, but in a Euclidean space. L_s, a_s, b_s axes are also closely aligned with L^*, a^*, b^*.

The space formulation is very complex, but nevertheless is promising for small color difference predictions. It highlights a known problem: the tradeoff between the accuracy of small and large differences.

2.8 SUMMARY AND NOTES

We have discussed color organization and modeling. More specifically, we have looked at various types of color organization models, such as process dependent, process order, pseudo perceptual, and perceptually uniform models. For more in depth discussions the reader is referred to the literature, e.g., [Lev88, Rob85, RO85, RO86].

3

COLOR IN COMPUTER GRAPHICS

In this Chapter[1] we discuss color organization and modeling in computer graphics. After a brief review of the Red, Green, and Blue (RGB) and Lightness, Hue, and Saturation (LHS) color models in computer graphics, we introduce, derive, and discuss a generalization of the latter, the Generalized Lightness, Hue, and Saturation (GLHS) model. We show that previously-used LHS color models are special cases of GLHS, and can be obtained from it by appropriate assignments to its free parameters. We derive some mathematical results concerning the relation between GLHS and RGB. Using these, we are able to give a single pair of simple algorithms for transforming from GLHS to RGB and vice versa. This single pair of algorithms transforms between RGB and any of the previously-published HSL, HSV, and HLS models, as well as any other special case of the generalized model. Nevertheless, they are as simple as the separate algorithms published previously. We give illustrations of color gamuts defined by various assignments to the free parameters of the GLHS system as they appear on the display monitor under the control of the Multiple Color Model Image Display System. Finally, we briefly discuss the potential for finding within the GLHS family a model that provides the closest approximation to a uniform color space. Such a model would share the perceptual properties of a proven uniform model and, at the same time, the algorithmic properties of the GLHS family. In Chapter 4, we describe a general optimization approach to finding such approximations, and two specific applications of this approach, to find closest approximations of the CIELUV uniform color space and the Munsell Book of Color, see also [LH88, LX92, LH93].

[1] This Chapter is adapted from [LH93].

3.1 INTRODUCTION

In computer graphics there is a need to specify colors in a way that is compatible with the hardware used (so it can be easily implemented) and at the same time is comprehensible to the user (so specification and recognition of colors and their components is manageable). These two requirements can be referred to as *hardware-* and *user-oriented* requirements, respectively. Unfortunately, it is difficult to find a model that fulfills both requirements.

We now discuss those color models that are most commonly used in computer graphics, namely the Red, Green, and Blue (RGB) model, which is used to model CRT color monitors, and the Lightness, Hue, and Saturation (LHS) family of models, which are considered to be better suited for human interaction. Because of the particular focus of this book on computer generated displays we use as our frame of reference the RGB model. Previously-developed LHS models can be perceived as having been derived by coordinate transformations of the RGB colorcube [LH87a]. Carrying that perception to its logical conclusion, a generalization is introduced, in the form of the Generalized Lightness, Hue, and Saturation model. In this generalization, three numbers (called *weights*) are used to select a specific model within the family. The previously-used models, as well as many new ones, are selected by specifying different values of the weights.

Section 3.2 discusses the color monitor and the RGB Color Cube, which models it. Section 3.3 discusses the family of LHS models that have been in use in computer graphics. These include the HSL-triangle and the HSV-hexcone developed by Smith [Smi78a, Smi78b] and Tektronix' HLS-double-hexcone. Our own slightly modified triangle model, the LHS-triangle, in which all specified colors are realizable on a monitor (which is not the case in the original triangle model), is also discussed. These lead us to our generalization, the Generalized Lightness, Hue, and Saturation (GLHS) model (see Section 3.4), which is defined in Section 3.4.1 in such a way that the previously used LHS models are special cases of GLHS. We also derive some basic mathematical properties of GLHS, especially of its relationship to RGB. Such properties allow us to devise surprisingly simple algorithms (as compared to those presented in [Lev88], for instance) for transforming between GLHS and RGB; these are described in Section 3.4.2, [LH93]. Section 3.5 illustrates the implementation and use of the GLHS model. The illustration of its potential use for a particular application, namely, the generation of integrated displays of multiparameter distributions is provided in Chapter 9. Finally, in Section 3.6, we discuss the advantages and disadvantages of the new model, and provide some directions for possible further research. Specifically, we discuss briefly the potential for finding within the GLHS family a model that provides the closest approximation to a uniform color space. Such a model would share the perceptual properties of a proven uniform model and, at the same time, the algorithmic properties of the GLHS family. In Chapter 4 we describe the details of such approximations.

3.2 THE COLOR MONITOR, THE COLORCUBE, AND THE RGB MODEL

Color display monitors create different colors by additive mixtures of the three primaries Red (R), Green (G), and Blue (B). For each primary there is an electron gun and a corresponding color-emitting phosphor on the screen surface. The intensity of each of the guns can be controlled (almost) independently between zero and a maximum voltage. Independent inputs for R, G, and B are used to control the guns and, thus, the color displayed on the screen.

Since the three guns are assumed to be varied independently and must have non-negative values less than or equal to a given maximum voltage, the gamut of an RGB color monitor can be represented by a cube, where the minimum gun voltage is represented by 0 and the maximum gun voltage by M.[2] This gamut is usually referred to as the *RGB Color Cube*, or just the *colorcube*, see Figure 3.1. Mathematically speaking, the colorcube consists of all points (r, g, b), such that $0 \leq r \leq M$, $0 \leq g \leq M$, and $0 \leq b \leq M$.[3]

A color in the colorcube model is a three-dimensional vector whose coordinates specify the amounts of r, g, and b that create it on the screen.[4] Although not every color that exists can be mixed by using non-negative amounts of red, green, and blue [Fis83], the gamut is large enough to be sufficient for most practical purposes.

Note that the RGB model does not provide a standard for exact color specification, since the color produced by a particular RGB specification depends on the spectral distribution of the primaries and the gamma characteristics of the display [Fis83]. The relationship between a typical RGB gamut and the collection of all existing colors can be seen in the *CIE Chromaticity Diagram*; see, e.g., [FvFH90] p. 585 and Color Plate II.2.

[2] Note that M need not be the same for each gun, though it usually is.

[3] Mathematical gamuts are defined continuously, which correspond to continuous gun voltages. However, digital inputs limit gamuts to discrete points.

[4] We use upper case letters for primaries (e.g., R, G, B) and lower case letters for their actual amounts (e.g., (r, g, b) means r amount of R, g amount of G, and b amount of B).

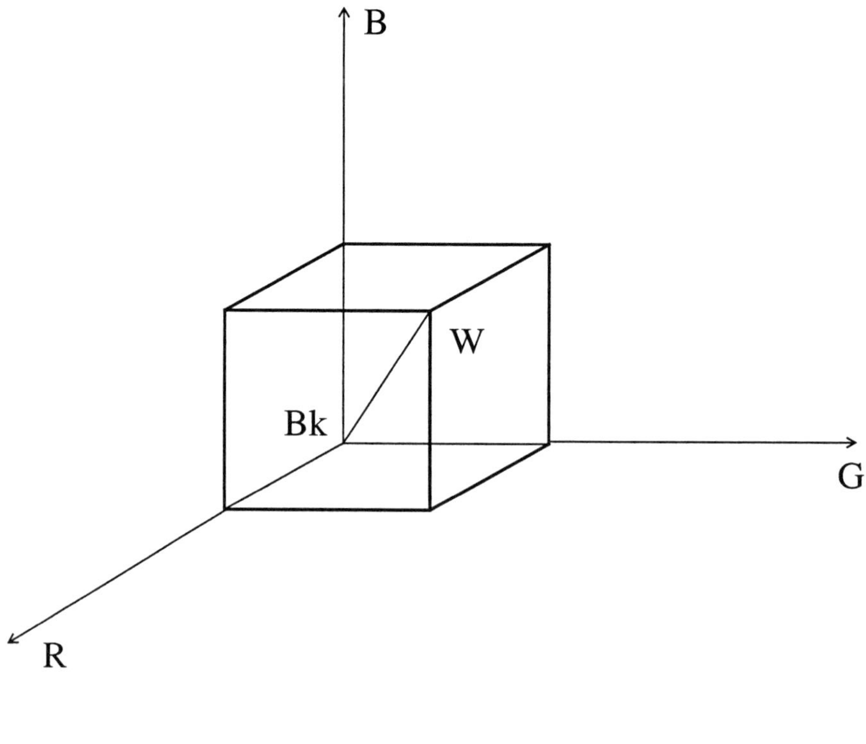

Figure 3.1 The RGB Colorcube.

3.3 THE LIGHTNESS, HUE, AND SATURATION (LHS) FAMILY OF MODELS

The colorcube model specifies colors in a straightforward way for display, but lacks an intuitive appeal. For example, it is not easy for the user to specify the exact amounts of red, green, and blue necessary to represent a dark brown. Similarly, in most cases it is difficult (and sometimes impossible) to estimate the amounts of the three primaries that were mixed to generate a color on the screen [Fis83, WC90].

An easier way to make such estimates corresponds to the way artists mix and specify colors using *tints*, *shades*, and *tones*, see Figure 3.2. An artist picks up a pure hue, mixes white in it to get a tint, mixes black in it to get a shade, or mixes both white and black to get a tone. The more white mixed in the color, the less saturated it will be, the more black in it, the darker (less light) it will be. The *LHS (lightness, hue, saturation)* models used in computer graphics, which are discussed below, are based on this notion of color mixing. To relate their three coordinates with the mixing notions of tints, shades, and tones (Figure 3.2), one has to note that decreasing saturation is "tinting" (corresponding to increasing whiteness) and decreasing lightness is "shading" (corresponding to increasing blackness). "Toning" is the combination of both.

Whenever a color space other than RGB is used, it is necessary to transform the color coordinates to RGB for display, and vice versa for color manipulation within the selected space. We now briefly describe several LHS models. More detailed descriptions of these models and of the way they are derived from the colorcube can be found in [Lev88]. We do not repeat that discussion here, since most of it is subsumed by the discussion of our general approach in Section 3.4.

All the models discussed in the following are derived from the colorcube by coordinate transformations from RGB to coordinates that represent the perceptual characteristics of color: lightness, hue, and saturation.[5] Roughly speaking, in all of them:

1. Approximate cylindrical coordinates are used. In this coordinate system (which is specified by the details that follow), the "polar" coordinates

[5] The names of the models were given at different times by different people, and thus they vary in the terms used for lightness and the order of components in the name. Synonymous terms can be found, e.g., in [Lev88]. Also note that the terms lightness and saturation have been defined less colloquially by color scientists. For such definitions, see, e.g., [Hun77].

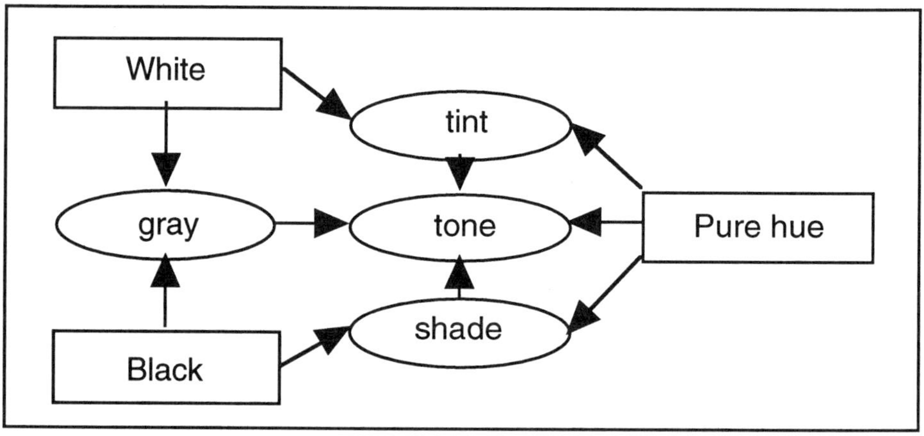

Figure 3.2 Tints, shades, and tones.

are the saturation s (proportional to radial distance) and the hue h (a function of the angle), with the lightness ℓ being the distance along the axis perpendicular to the polar coordinate plane.

2. In the transformation from (r, g, b) to (ℓ, h, s), exactly those points in the colorcube for which $r = g = b$ are assigned zero saturation ($s = 0$), and hence undefined hue h. (These colors are the grays, also referred to as *achromatic colors.*) Furthermore, for these colors, the lightness ℓ is given the common value of r, g, and b. Geometrically, we can picture this by considering the colorcube being stood on its vertex indicated by **Bk** in Figure 3.1 (the black point) with the main diagonal of the cube from **Bk** to **W** (the white point) corresponding to the positive lightness axis from 0 to M.

3. The lightness ℓ assigned to an arbitrary point (r, g, b) in the colorcube is defined in such a way that:
   (a) the value of lightness is always between 0 and M, and
   (b) the set of points (r, g, b) in the colorcube that are assigned a common value of ℓ form a *constant-lightness surface* with a special property, namely that any line parallel to the main diagonal of the colorcube meets the surface at no more than one point. (The members of the LHS family differ from each other in the actual shapes of these surfaces. Since we restrict these surfaces to be subsets of the colorcube, we will have a few pathological cases in which a "surface" contains a single point or is the union of three line segments.)

In the process of transforming the colorcube into a color solid of the LHS family, every one of the constant-lightness surfaces, defined in 3b above, is projected onto a plane perpendicular to the lightness axis intersecting it at the origin. The projection of the constant-lightness surface onto this plane defines a shape (e.g., a triangle, a hexagonal disk, etc.) that depends on the lightness function chosen and the specific lightness value. The projected constant-lightness surface is then moved back so that it intersects the lightness axis at its lightness value. Repeating the process for all lightness values, stacks all the projected constant-lightness surfaces in the order of their lightnesses ($\ell = 0$ at the bottom, $\ell = M$ at the top). This yields a three-dimensional body, which is the color solid for the particular model. Note that since the entire process of projecting constant-lightness surfaces and color vectors is done in RGB space, the shape of the resulting color solid varies as a function of the lightness function.

In what follows, we use the phrase *projected color vector* of a color (r, g, b) to mean the projection of (r, g, b) onto the plane through the origin (the black point) perpendicular to the lightness axis. Mathematically, the projected color vector of (r, g, b) is the vector $(\frac{2r-g-b}{3}, \frac{2g-b-r}{3}, \frac{2b-r-g}{3})$. This implies that the location in the color solid of the point that corresponds to (r, g, b) in the colorcube is, in (r, g, b)-coordinates, $(\frac{2r-g-b}{3} + \ell, \frac{2g-b-r}{3} + \ell, \frac{2b-r-g}{3} + \ell)$, where ℓ is the lightness of (r, g, b). So for the LHS family under discussion, the shape of the color solid (and the transformation of the colorcube into it) depends only on the definition of lightness (and not on the definition of hue and saturation).

4. The hue h of a chromatic color (r, g, b) is defined by a function of the angle between its projected color vector and a predefined vector (traditionally the projected color vector of a pure red). Typically, this function is chosen so that:

   (a) it maps 0 into 0 and its whole domain $[0, 360)$ onto $[0, 360)$;

   (b) it is continuous and monotonically increasing; and

   (c) its value for any argument is an approximation of that argument.

   We point out here the important property that the angle between the projected color vectors of any two chromatic colors is independent of the particular choice of the lightness function. Hence, in all LHS models in which the same function satisfying the conditions above is used to specify hue, the hue assigned to a particular color (r, g, b) will be the same. Furthermore, for the same reason, the hue of a color (r, g, b) will be unchanged by the addition or subtraction of an achromatic color (i.e., by the tinting and shading processes of Figure 3.2). This is a valuable property for some applications, such as integrated visualization of multiparameter distributions, discussed in Chapter 9.

5. The saturation s of a color (r, g, b) is defined as the ratio of the length of its projected color vector to the length of the longest projected color vector in the same direction, in the same constant-lightness surface. Thus, for the vectors of any fixed constant-lightness surface, a color that has the longest projected color vector (in any particular direction) has maximum saturation ($s = 1$) and the achromatic color has minimum saturation ($s = 0$).

The essential choice in selecting a particular LHS model is made in the definition of the lightness function, which in turn determines the constant-lightness surfaces (and hence the shape of the color solid that represents the model). An independent secondary choice is made in selecting the particular function satisfying 4a–c in the definition of hue. Once the lightness function is chosen, saturation is completely defined by 5. (In particular, it does not depend on the choice of the function used in the definition of hue.)

The LHS-triangle model

The simplest way to choose constant-lightness surfaces is to define them as planes. The triangle model defines the lightness $\ell(\mathbf{c})$ of a color $\mathbf{c} = (r, g, b)$ as:

$$\ell(\mathbf{c}) = \frac{r+g+b}{3}, \tag{3.1}$$

where the division by 3 serves only to normalize the lightness into the range $[0, M]$. A constant-lightness surface with lightness ℓ is the plane:

$$\{(r,g,b) : (r+g+b)/3 = \ell\}. \tag{3.2}$$

For $0 \leq \ell \leq M$, these planes are perpendicular to the main diagonal of the colorcube and parallel to each other. Thus, in this case, the constant-lightness surfaces are "projected" onto themselves and so the color solid is still cubic in shape. The shape of a constant lightness surface is a triangle for $0 \leq \ell \leq M/3$ and for $2M/3 \leq \ell \leq M$, and is a hexagon for values of ℓ in between. The LHS-triangle model has been introduced in [LH87a] as a variant of the inxHSL-triangle model of Smith [Smi78a]. In the latter model every constant-lightness surface is a triangle, but the set of colors in the model is not exactly the set of colors realizable on an RGB color monitor. The colors in the LHS-triangle model are (by definition) exactly those in the colorcube. Algorithms for transforming from (r, g, b) coordinates to (ℓ, h, s) coordinates and the other way around have been described in [Lev88]. These algorithms can be replaced by the simpler ones that we provide below to transform coordinates in either direction between the RGB model and the GLHS model, because the LHS-triangle model is a special case of the GLHS model.

The HSV-hexcone model

The HSV-hexcone model can be derived from the colorcube by defining a different lightness function and hence different constant-lightness surfaces. Here, the lightness (which is typically called *value* by users of this model) of a given color $\mathbf{c} = (r, g, b)$ is calculated as:

$$\ell(\mathbf{c}) = max\{r, g, b\}. \tag{3.3}$$

The locus of all the colors with value ℓ are the points:

$$\{(r,g,b) : max\{r,g,b\} = \ell\}\,. \tag{3.4}$$

Limiting the coordinates so that $0 \leq r,g,b \leq M$ yields the surface that consists of the three faces "visible from above" of the subcube (originating at the origin of the colorcube) whose side length equals ℓ. We call this subcube the *ℓ-subcube*. (Its corner opposite **Bk** is the vector (ℓ,ℓ,ℓ).) Projecting the visible faces along the main diagonal onto a plane perpendicular to the diagonal yields a hexagonal disk. In the disk, the R, G, and B axes are 120 degrees apart from each other, and complementary colors (e.g., cyan and red) are 180 degrees opposite one another. Each sector of the disk corresponds to one half of a visible face divided along the face diagonal that connects the point (ℓ,ℓ,ℓ) with the opposite corner of the face. For each ℓ-subcube there is a hexagonal disk corresponding to it. Stacking the disks vertically, with the smallest disk at the bottom (a point—corresponding to the origin—which is the black point) and the largest at the top (corresponding to the entire colorcube) yields a hexagonal cone (*hexcone*). On the envelope of the hexcone lie all the fully saturated colors (with saturation equal to 1, i.e., primaries and colors that are mixtures of at most two primaries). Moving in towards the central axis of the hexcone corresponds to desaturating ("tinting"); with the extreme on the central axis where colors have 0 saturation, that is, they are achromatic colors. The complete derivation of the model and the transformation algorithms between the RGB and the HSV models are given in [Smi78a]. Additional discussions can be found in [FvFH90, JG78].

The HLS-double-hexcone model

In this model, the lightness $\ell(\mathbf{c})$ of a color $\mathbf{c} = (r,g,b)$ is defined as:

$$\ell(\mathbf{c}) = \frac{max\{r,g,b\} + min\{r,g,b\}}{2}. \tag{3.5}$$

This lightness function gives rise to a set of ℓ-subcubes, such that the main diagonal of each of them is a segment of the main diagonal of the colorcube with the following starting and ending points:

$$\begin{cases} (0,0,0) & \rightarrow \quad (2\ell,2\ell,2\ell), \quad \text{if } 0 \leq \ell \leq M/2, \\ (2\ell - M, 2\ell - M, 2\ell - M) & \rightarrow \quad (M,M,M), \quad \text{if } M/2 < \ell \leq M. \end{cases} \tag{3.6}$$

For each ℓ, the constant-lightness surface is the locus of points

$$\{(r,g,b) : max\{r,g,b\} + min\{r,g,b\} = 2\ell\}\,. \tag{3.7}$$

	pure hues	secondary	white
HSL-Triangle	$M/3$	$2M1/3$	M
HSV-Hexcone	M	M	M
HLS-Double-Hexcone	$M/2$	$M/2$	M

Table 3.1 Comparison of the lightness of pure hues, secondary colors, and white in the three LHS models.

Limiting the color coordinates to $0 \leq r, g, b \leq M$ yields a surface that consists of the six triangles resulting from connecting the point (ℓ, ℓ, ℓ)—located in the middle of the main diagonal of the ℓ-subcube—to the six corners of the subcube that are not connected by the main diagonal. Projecting these triangles along the main diagonal onto a plane perpendicular to it again yields a hexagonal disk, similar to the one in the HSV-hexcone. The only differences are that the largest disk corresponds now to $\ell = M/2$ and the disks for both $\ell = 0$ and $\ell = M$ (the black and white points) are single points. Stacking the disks vertically in the order of the values of ℓ that they represent yields a double hexagonal cone (*double-hexcone*) with primaries located on the largest disk ($\ell = M/2$) in the same way they are organized in the HSV-hexcone model. Thus, in the HLS-double-hexcone the pure hues have lightness $M/2$ while in the HSV-hexcone their lightness is M.

Table 3.1 summarizes the differences among the lightnesses of pure hues (primaries), secondary colors, and white in the three LHS colors. Note that the property of the triangle model—where the lightness for each one of these groups differs—provides an advantage in applications that use both color and monochrome displays: the three groups of colors can be distinguished on a monochrome display based on their lightness alone (compare to the hexcone model, where they are the same) .

3.4 GLHS: A GENERALIZED LIGHTNESS, HUE, AND SATURATION MODEL

3.4.1 Definition and basic properties

The various LHS models presented in the previous section have common properties and are derived from the colorcube using similar concepts. This gives rise to a generalization that describes a whole class of models. GLHS, the generalized lightness, hue, and saturation color model, provides a first-order mathematical framework for that class of models. The models described above are special cases of GLHS, realized by special parameter values [Lev88, LH87a, LH93].

This first-order generalization uses piecewise planar constant-lightness surfaces. We define three non-negative weights w_{min}, w_{mid}, w_{max}, such that $w_{max} > 0$ and $w_{min} + w_{mid} + w_{max} = 1$. The lightness function is defined as:

$$\ell(\mathbf{c}) = w_{min} \cdot min(\mathbf{c}) + w_{mid} \cdot mid(\mathbf{c}) + w_{max} \cdot max(\mathbf{c}), \tag{3.8}$$

where $min(\mathbf{c})$, $mid(\mathbf{c})$, and $max(\mathbf{c})$ are defined as:

$$\begin{aligned} min(\mathbf{c}) &= min\{r, g, b\}, \\ mid(\mathbf{c}) &= mid\{r, g, b\}, \\ max(\mathbf{c}) &= max\{r, g, b\}, \end{aligned} \tag{3.9}$$

and a constant-lightness surface for a given lightness ℓ is given by the locus of points:

$$\{\mathbf{c} : w_{min} \cdot min(\mathbf{c}) + w_{mid} \cdot mid(\mathbf{c}) + w_{max} \cdot max(\mathbf{c}) = \ell\}. \tag{3.10}$$

Generally, this consists of the six planar polygons corresponding to the six combinations of the order of the magnitudes of r, g, and b. Pathological cases arise when some of the six planes intersect the colorcube in a point or a line. For example, since $w_{max} > 0$, this is the case in all GLHS models where the only color $\mathbf{c}$ for which $\ell(\mathbf{c}) = 0$ is the black point. In the special case when $w_{max} = w_{mid} = 1/2$, the set of colors for which $\ell(\mathbf{c}) = M$ form the three edges of the colorcube that meet at the white point. In the discussion that follows we have made sure that the mathematics is valid for the pathological cases as well as for the general case. In particular, it follows from the properties of the weights and from the ranges of the primaries that for all GLHS models $0 \leq \ell(\mathbf{c}) \leq M$, for any color $\mathbf{c}$.

Different values of w_{min}, w_{mid}, and w_{max} give rise to different color models. Table 3.2 gives the values of the weights for the models discussed in Section 3.3. By changing the values of the three weights, a continuum of models can be achieved. To complete the definition of a GLHS model, we need to describe how, for a color $\mathbf{c}$, the hue $h(\mathbf{c})$ and the saturation $s(\mathbf{c})$ are defined.

The hue $h(\mathbf{c})$ of a chromatic color $\mathbf{c} = (r, g, b)$, $0 \leq h(\mathbf{c}) < 360$, is defined as follows:

$$h(\mathbf{c}) = (k(\mathbf{c}) + f(\mathbf{c})) \cdot 60, \tag{3.11}$$

	w_{min}	w_{mid}	w_{max}
HSL-Triangle	1/3	1/3	1/3
HSV-Hexcone	0	0	1
HLS-Double-Hexcone	1/2	0	1/2

Table 3.2 The values of the three weights that realize the computer-graphics color models.

where $k(\mathbf{c}) \in \{0, 1, \ldots 5\}$ is the number of the *sector* defined by the order of the magnitudes of the r, g, and b values:

$$k(\mathbf{c}) = \begin{cases} 0, & \text{if } r > g \geq b, \\ 1, & \text{if } g \geq r > b, \\ 2, & \text{if } g > b \geq r, \\ 3, & \text{if } b \geq g > r, \\ 4, & \text{if } b > r \geq g, \\ 5, & \text{if } r \geq b > g, \end{cases} \tag{3.12}$$

and $f(\mathbf{c}) \in [0, 1)$, the *hue-fraction* within the sector, is calculated as follows:

$$f(\mathbf{c}) = \begin{cases} \frac{mid(\mathbf{c}) - min(\mathbf{c})}{max(\mathbf{c}) - min(\mathbf{c})}, & \text{if } k(\mathbf{c}) \text{ is even}, \\ \frac{max(\mathbf{c}) - mid(\mathbf{c})}{max(\mathbf{c}) - min(\mathbf{c})}, & \text{if } k(\mathbf{c}) \text{ is odd}. \end{cases} \tag{3.13}$$

This is a modified representation of one of the hue functions presented in [Smi78a]. It satisfies all the properties specified for hue in item 4 of Section 3.3; for a proof see Section 3.5.2.[6] In particular, we show in Section 3.5.2 that $h(\mathbf{c})$ as defined by Equation 3.11 approximates the angle between the projected color vector of the color in question and that of the red axis extremely well, with the difference between the two values being never more than 1.12 degrees. We prefer our definition to the use of the angle itself, since it does away with the need for trigonometric functions in the transformations to and from the RGB model. One can even argue that it is the more natural definition, since the $f(\mathbf{c})$ function directly provides the ratio of the lesser component to the larger component in the "pure hue" associated with that color after all the desaturating white has been removed (see Figure 3.2).

[6] We leave this proof to the end to maintain the flow of the discussion, and to allow those readers who are not interested in such mathematical detail to skip it.

Note that for a chromatic color, $max(\mathbf{c}) > min(\mathbf{c})$ and so $f(\mathbf{c})$ (and thus $h(\mathbf{c})$) is well defined. Furthermore, the definition of hue is independent of the definition of lightness; for any chromatic color $\mathbf{c}$, the value $h(\mathbf{c})$ of the hue of $\mathbf{c}$ is the same in all the GLHS models.

The saturation $s(\mathbf{c})$ of a color $\mathbf{c} = (r, g, b)$ is completely defined by the description in item 5 of Section 3.3. We now derive a mathematical expression for $s(\mathbf{c})$ in the GLHS model.

Since in each hue-sector the constant-lightness surface is planar, for any color $\mathbf{c}$, the set of colors $\mathbf{d}$ that are in the same constant-lightness surface as $\mathbf{c}$, and whose projected color vectors are in the same direction as that of $\mathbf{c}$ is exactly the set of colors of the form $\mathbf{d} = \mathbf{L}(\mathbf{c}) + t(\mathbf{c} - \mathbf{L}(\mathbf{c}))$, where $t > 0$ and $\mathbf{L}(\mathbf{c}) = (\ell(\mathbf{c}), \ell(\mathbf{c}), \ell(\mathbf{c}))$ is the achromatic point in the constant-lightness surface of $\mathbf{c}$ (see Figure 3.3). Therefore, the saturation of any color $\mathbf{c}$ is equal to the ratio of the length of the vector connecting $\mathbf{L}(\mathbf{c})$ to $\mathbf{c}$ to the length of the extension of this vector to the surface of the colorcube. Thus, $s(\mathbf{c})$ can be defined as the inverse of the maximum value of t such that $\mathbf{L}(\mathbf{c}) + t(\mathbf{c} - \mathbf{L}(\mathbf{c}))$ is in the colorcube. (This gives zero if $\mathbf{c} = \mathbf{L}(\mathbf{c})$.)

For a chromatic color $\mathbf{c}$, there are two cases for the computation of $s(\mathbf{c})$, depending on whether the extension of the vector from $\mathbf{L}(\mathbf{c})$ to $\mathbf{c}$ intersects the surface of the colorcube at a face whose colors have a minimum coordinate 0 (attached to black) or at a face whose colors have a maximum coordinate M (attached to white). To help us make this distinction we define a color $\mathbf{q}(\mathbf{c})$, which depends on $\mathbf{c}$ as follows:

$$\mathbf{q}(\mathbf{c}) = \begin{cases} (M, f(\mathbf{c})M, 0), & \text{if } k(\mathbf{c}) = 0, \\ ((1 - f(\mathbf{c}))M, M, 0), & \text{if } k(\mathbf{c}) = 1, \\ (0, M, f(\mathbf{c})M), & \text{if } k(\mathbf{c}) = 2, \\ (0, (1 - f(\mathbf{c}))M, M), & \text{if } k(\mathbf{c}) = 3, \\ (f(\mathbf{c})M, 0, M), & \text{if } k(\mathbf{c}) = 4, \\ (M, 0, (1 - f(\mathbf{c}))M), & \text{if } k(\mathbf{c}) = 5. \end{cases} \tag{3.14}$$

Note that $k(\mathbf{q}(\mathbf{c})) = k(\mathbf{c})$, $f(\mathbf{q}(\mathbf{c})) = f(\mathbf{c})$, and so $h(\mathbf{q}(\mathbf{c})) = h(\mathbf{c})$. In fact $\mathbf{q}(\mathbf{c})$ depends only on $f(\mathbf{c})$ and $k(\mathbf{c})$ and thus it is the same for all colors of the same hue.

Since $min(\mathbf{q}(\mathbf{c})) = 0$ and $max(\mathbf{q}(\mathbf{c})) = M$, $\mathbf{q}(\mathbf{c})$ occupies a special place in the triangle in the colorcube containing all points of the same hue as $\mathbf{c}$; namely, it is the third vertex. (The other two are **Bk** and **W**, see Figure 3.3.) Since in the GLHS models the intersections of constant-lightness surfaces with this triangle are all parallel to each other, we see that the extension of the vector

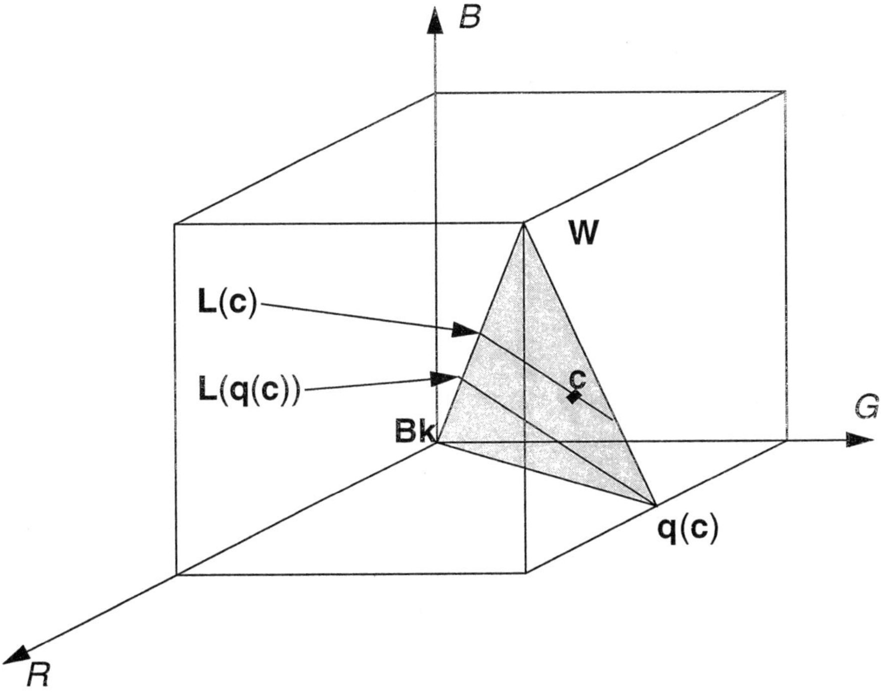

Figure 3.3 The triangle in the colorcube containing all points of the same hue as c.

from $\mathbf{L}(\mathbf{c})$ to $\mathbf{c}$ intersects the surface of the colorcube at the face whose colors have minimum coordinate zero if and only if $\ell(\mathbf{c}) \leq \ell(\mathbf{q}(\mathbf{c}))$. Since, for all positive t,

$$min(\mathbf{L}(\mathbf{c}) + t(\mathbf{c} - \mathbf{L}(\mathbf{c}))) = \ell(\mathbf{c}) + t(min(\mathbf{c}) - \ell(\mathbf{c})), \tag{3.15}$$

the largest value of t for which $\mathbf{L}(\mathbf{c}) + t(\mathbf{c} - \mathbf{L}(\mathbf{c}))$ is still in the colorcube is, in this case,

$$\frac{\ell(\mathbf{c})}{\ell(\mathbf{c}) - min(\mathbf{c})}. \tag{3.16}$$

By a similar argument in the case where $\ell(\mathbf{c}) > \ell(\mathbf{q}(\mathbf{c}))$, and by the previously stated property of $s(\mathbf{c})$, we get

$$s(\mathbf{c}) = \begin{cases} \frac{\ell(\mathbf{c}) - min(\mathbf{c})}{\ell(\mathbf{c})}, & \text{if } \ell(\mathbf{c}) \leq \ell(\mathbf{q}(\mathbf{c})), \\ \frac{max(\mathbf{c}) - \ell(\mathbf{c})}{M - \ell(\mathbf{c})}, & \text{if } \ell(\mathbf{c}) > \ell(\mathbf{q}(\mathbf{c})). \end{cases} \tag{3.17}$$

Note that the definition in Equation 3.17 is made easier by the fact that

$$\begin{aligned} \ell(\mathbf{q}(\mathbf{c})) &= w_{mid} f'(\mathbf{c}) M + w_{max} M & (3.18) \\ &= w_{mid} \frac{mid(\mathbf{c}) - min(\mathbf{c})}{max(\mathbf{c}) - min(\mathbf{c})} M + w_{max} M, & (3.19) \end{aligned}$$

where $f'(\mathbf{c})$ is defined to be

$$f'(\mathbf{c}) = \begin{cases} f(\mathbf{c}), & \text{if } k(\mathbf{c}) \text{ is even}, \\ 1 - f(\mathbf{c}), & \text{if } k(\mathbf{c}) \text{ is odd}, \end{cases} \tag{3.20}$$

as can be easily derived from Equations 3.8, 3.13, and 3.14.

Note that even though at first sight it appears that there is a potential for division by zero in Equation 3.17, this is in fact not the case for a chromatic color, as we now show. Since $w_{max} > 0$, by Equation 3.8 $\ell(\mathbf{c}) = 0$ only if $max(\mathbf{c}) = 0$, in which case $\mathbf{c}$ is the black point. Similarly, if $w_{min} > 0$, then $\ell(\mathbf{c}) = M$ only if $min(\mathbf{c}) = M$, in which case $\mathbf{c}$ is the white point. On the other hand, $w_{min} = 0$ and $\ell(\mathbf{c}) = M$ together imply that $\ell(\mathbf{q}(\mathbf{c})) = M$. (If $mid(\mathbf{c}) = max(\mathbf{c})$, then this follows from Equation 3.19 straight away. If $mid(\mathbf{c}) < max(\mathbf{c})$, then by Equation 3.8 we must have $w_{mid} = 0$ and $w_{max} = 1$, and again Equation 3.19 provides the desired result.) For $\ell(\mathbf{q}(\mathbf{c})) = M$, the first line of Equation 3.17 is operative, and so $\ell(\mathbf{c}) = M$ does not cause a division by zero.

Since the achromatic point in the constant-lightness surface of a color depends not only on the color but also on the choices of w_{min}, w_{mid}, and w_{max}, the saturation of a color depends on the particular model chosen from the GLHS family. This is, of course, also true for the lightness of the color (see Equation 3.8), but not for the hue, as it is clear from the definition given above (as well as from the general discussion in item 4 of Section 3.3).

We now discuss the range of the mapping from RGB-coordinates to LHS-coordinates. For any achromatic color $\mathbf{c} = (\ell, \ell, \ell)$, $0 \leq \ell \leq M$, we have $\ell(\mathbf{c}) = \ell$, $h(\mathbf{c})$ is undefined, and $s(\mathbf{c}) = 0$. So the set of achromatic colors maps onto the set $R_A = [0, M] \times \{undefined\} \times \{0\}$. This mapping is one-to-one, with different achromatic colors mapping into different elements of R_A. Now consider the set of chromatic colors $\mathbf{c}$ for which $0 < \ell(\mathbf{c}) < M$. Clearly (see Figure 3.3), this set maps onto the set $R_C = (0, M) \times [0, 360) \times (0, 1]$. Again, different chromatic colors with $0 < \ell(\mathbf{c}) < M$ do not map into the same element of R_C. (If $h(\mathbf{c_1}) = h(\mathbf{c_2})$, then $\mathbf{c_1}$ and $\mathbf{c_2}$ lie on the same plane through the achromatic axis; if also $\ell(\mathbf{c_1}) = \ell(\mathbf{c_2})$, then they lie on the same line through $\mathbf{L}(\mathbf{c_1}) = \mathbf{L}(\mathbf{c_2})$; and if also $s(\mathbf{c_1}) = s(\mathbf{c_2})$, then they must be the same point on this line; see Figure 3.3.) Since there is no chromatic color $\mathbf{c}$ with $\ell(\mathbf{c}) = 0$, the only colors whose images have not yet been discussed are the set of chromatic colors $\mathbf{c}$ with $\ell(\mathbf{c}) = M$. Let R_B denote the range onto which these colors are mapped. We need to distinguish between three cases. If $w_{min} > 0$, then there is no chromatic color $\mathbf{c}$ such that $\ell(\mathbf{c}) = M$ and so R_B is the empty set. If $w_{min} = 0$ and $w_{mid} > 0$, then $\ell(\mathbf{c}) = M$ implies that $max(\mathbf{c}) = mid(\mathbf{c}) = M$, while $min(\mathbf{c})$ can be anything. These colors lie on the three edges of the colorcube which meet at the white point $\mathbf{W}$. In such a case, the constant-lightness surfaces are parallel to the axis associated with the coordinate of $min(\mathbf{c})$. Looking at Figure 3.3, we see that for any $\mathbf{c}$ with $max(\mathbf{c}) = mid(\mathbf{c}) = M$ and $min(\mathbf{c}) < M$, we have $\mathbf{L}(\mathbf{c}) = \mathbf{L}(\mathbf{q}(\mathbf{c})) = \mathbf{W}$, and $s(\mathbf{c})$ is in the range $(0, 1]$. Thus, in the case $w_{min} = 0$ and $w_{mid} > 0$, $R_B = \{M\} \times \{60, 180, 300\} \times (0, 1]$. Again, the mapping is one-to-one. Finally, if $w_{min} = w_{mid} = 0$, then $w_{max} = 1$. This is the HSV-hexcone case. As discussed before, in this case the set of chromatic colors $\mathbf{c}$ for which $\ell(\mathbf{c}) = M$ are on the three faces of the colorcube that meet at the white point. Here again we have that $\mathbf{L}(\mathbf{c}) = \mathbf{L}(\mathbf{q}(\mathbf{c})) = \mathbf{W}$ and $s(\mathbf{c})$ is in the range $(0, 1]$, but this time the hue is not restricted to three specific directions but may assume any of its values. Thus, $R_B = \{M\} \times [0, 360) \times (0, 1]$. The mapping onto R_B is one-to-one. Note that the intersections $R_A \cap R_B$, $R_A \cap R_C$, and $R_B \cap R_C$ are all empty.

In summary, the transformation from RGB-coordinates to LHS-coordinates is a one-to-one mapping onto its range $R_A \cup R_B \cup R_C$. It therefore has an inverse defined on this range. Combinations of (ℓ, h, s) coordinates that are not in $R_A \cup R_B \cup R_C$ do not correspond to any color in the colorcube.

Note that the generalization presented here affects the shape of the constant-lightness surfaces and the lightness ranges for which they hold and thus the shape of the color solid. The shapes and ranges are as follows (see Figures 3.4–3.6):

For $0 < \ell \leq M w_{max}$ and $M(w_{max} + w_{mid}) \leq \ell \leq M$: the projected constant-lightness surface's shape is generally either a triangle or a hexagon, depending on the values of the weights. (For example, it is a triangle for $0 < \ell \leq M/3$ and $2M/3 \leq \ell < M$ in the LHS-triangle model.)

For $M w_{max} < \ell < M(w_{max} + w_{mid})$: the projected constant-lightness surface's shape is generally either a hexagon or a dodecagon. (For example, it is a hexagon for $M/3 < \ell < 2M/3$ in the LHS-triangle model.)

For the HSV-hexcone, $w_{max} = 1$ and thus $0 < \ell \leq M w_{max}$ yields $0 < \ell \leq M$ as the only case. For the HLS-double-hexcone, $w_{mid} = 0$ and thus the only cases are $0 < \ell \leq M w_{max}$ and $M(w_{max} + w_{mid}) \leq \ell \leq M$, and these simplify to $0 < \ell \leq M/2$ and $M/2 < \ell \leq M$, respectively. For the LHS-triangle, all three cases apply. For the special case $w_{max} = w_{mid} = 1/2$, we have triangles for $0 < \ell < M/2$. Three pairs of edges become each a single edge in the dodecagon for $M/2 < \ell < M$, resulting in a (non-convex) polygon with nine edges. In the limiting case of $\ell = M$, we end up with the already discussed pathological extreme of an infinitely thin hexagonal star which consists of three line segments radiating from a common point (the white point).

3.4.2 Algorithms to transform to and from RGB

In this Section we present a pair of algorithms to handle transformations between the RGB and GLHS models. This pair is sufficient to transform between RGB and any member of the GLHS family.

RGB to GLHS

The algorithm presented in Figure 3.7 computes the (ℓ, h, s) values of a color $\mathbf{c} = (r, g, b)$ in some GLHS model defined by the weights w_{max}, w_{mid}, and w_{min}.

The case of an achromatic color is self-explanatory. If the color is chromatic, we compute its lightness $\ell(\mathbf{c})$ using Equation 3.8. The computation of the hue is done in three steps:

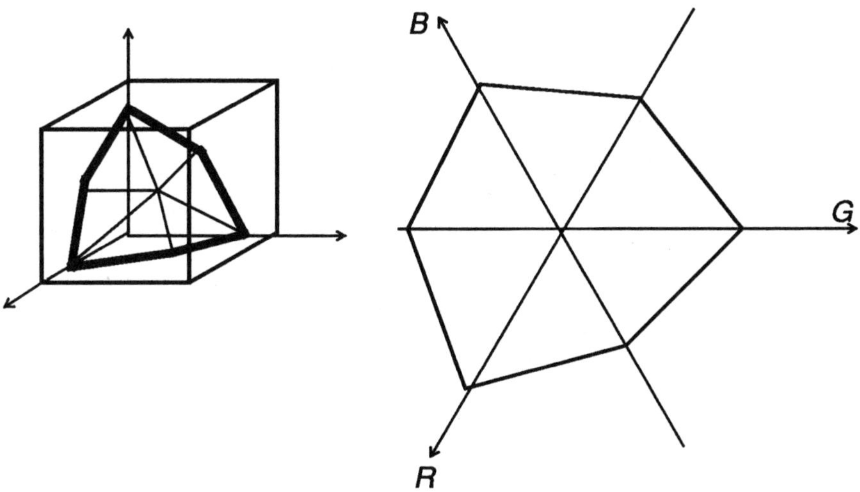

Figure 3.4 A general cross section through the GLHS Model: $0 < \ell \leq Mw_{max}$.

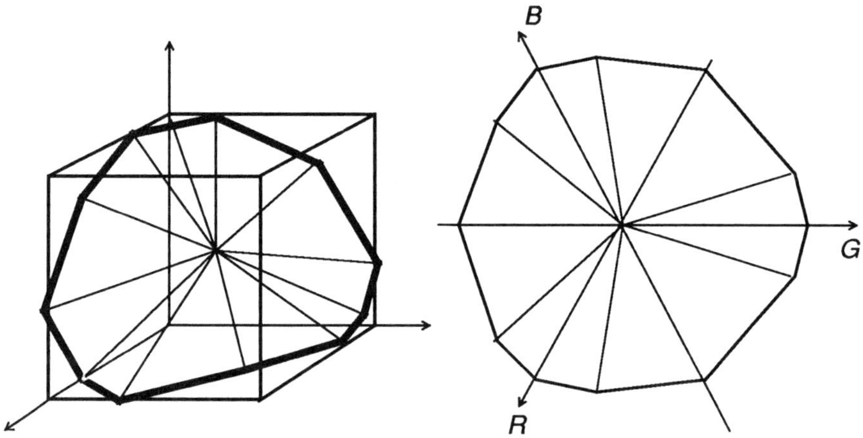

Figure 3.5 Another general cross section through the GLHS Model: $Mw_{max} < \ell < M(w_{max} + w_{mid})$.

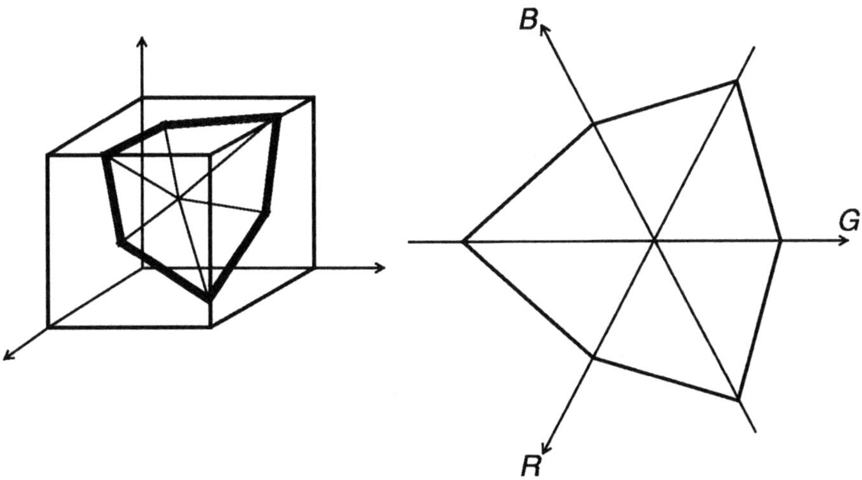

Figure 3.6 A third general cross section through the GLHS Model: $M(w_{max} + w_{mid}) \leq \ell \leq M$.

Algorithm *RGB_TO_GLHS*
Input: $\mathbf{c} = (r, g, b) \in [0, M]^3$,
$w_{max}, w_{mid}, w_{min}$, such that $0 \leq w_{max}, w_{mid}, w_{min} \leq 1$, $w_{max} > 0$,
and $w_{max} + w_{mid} + w_{min} = 1$.
Output: (ℓ, h, s), $\ell \in [0, M]$, $h \in [0, 360) \cup \{undefined\}$, $s \in [0, 1]$.
Auxiliary variables: the critical lightness $\ell(\mathbf{q})$, k, f.

```
begin
    max := MAXIMUM(r, g, b);
    mid := MID_VALUE(r, g, b);
    min := MINIMUM(r, g, b);
    if max = min
        then                                    {achromatic}
            (ℓ, h, s) := (max, undefined, 0)
        else begin                              {chromatic}
            ℓ := w_max * max + w_mid * mid + w_min * min;
            begin case of                       {sector-number k}
                r > g ≥ b : k := 0;
                g ≥ r > b : k := 1;
                g > b ≥ r : k := 2;
                b ≥ g > r : k := 3;
                b > r ≥ g : k := 4;
                r ≥ b > g : k := 5;
            endcase
            begin case of                       {hue-within-sector f}
                k even: f := (mid − min)/(max − min);
                k odd: f := (max − mid)/(max − min);
            endcase
            h := (k + f) * 60;
            ℓ(q) = (w_mid * (mid − min)/(max − min) + w_max) * M;
            if ℓ ≤ ℓ(q)
                then s := (ℓ − min)/ℓ;
                else s := (max − ℓ)/(M − ℓ);
        end                                     {chromatic}
end; {RGB_TO_GLHS}
```

Figure 3.7 The RGB-TO-GLHS transformation algorithm.

1. Compute the sector number $k(\mathbf{c})$ using Equation 3.12.
2. Compute the hue-fraction $f(\mathbf{c})$ within the sector $k(\mathbf{c})$ using Equation 3.13.
3. Compute $h(\mathbf{c})$ from $k(\mathbf{c})$ and $f(\mathbf{c})$ using Equation 3.11.

Now we compute the lightness $\ell(\mathbf{q}(\mathbf{c}))$ of the point $\mathbf{q}(\mathbf{c})$, using Equation 3.19. Finally, the saturation $s(\mathbf{c})$ is computed using Equation 3.17.

GLHS to RGB

The algorithm in Figure 3.8 computes the color $\mathbf{c} = (r, g, b)$ in RGB coordinates for a color (ℓ, h, s) given in some GLHS model defined by the weights w_{max}, w_{mid}, and w_{min}. Note that the input is restricted to those combinations of the (ℓ, h, s) coordinates that correspond to a color in the colorcube.

If $s = 0$, then $\mathbf{c} = (\ell, \ell, \ell)$.

In the chromatic case ($s \neq 0$), first we compute the sector $k(\mathbf{c})$ and the hue-fraction $f(\mathbf{c})$ within the sector. We then compute the value of $f'(\mathbf{c})$ using Equation 3.20. Now we can compute $\ell(\mathbf{q}(\mathbf{c}))$ using Equation 3.18. Next, we compute $min(\mathbf{c})$ (if $\ell \leq \ell(\mathbf{q}(\mathbf{c}))$) or $max(\mathbf{c})$ (if $\ell > \ell(\mathbf{q}(\mathbf{c}))$), from Equation 3.17. Now, we are ready to compute $mid(\mathbf{c})$ and either $max(\mathbf{c})$ or $min(\mathbf{c})$ (depending on which one we computed previously). The derivation of these expressions from Equations 3.8, 3.13, and 3.20 is straightforward. In all cases, we have to make sure that the denominator is not equal to zero. If $\ell \leq \ell(\mathbf{q}(\mathbf{c}))$, then this is clearly the case since $w_{max} > 0$. If $\ell > \ell(\mathbf{q}(\mathbf{c}))$, then substituting for $\ell(\mathbf{q}(\mathbf{c}))$ using Equation 3.18, we get that $w_{mid}(1 - f'(\mathbf{c})) + w_{min} > \frac{M-\ell}{M}$. Since $\frac{M-\ell}{M} \geq 0$ we get that $w_{mid}(1 - f'(\mathbf{c})) + w_{min} > 0$. This allows us to calculate $mid(\mathbf{c})$. If $w_{min} \neq 0$, we can now compute $min(\mathbf{c})$ directly from Equation 3.8. If $w_{min} = 0$, we get

$$min(\mathbf{c}) = \frac{mid(\mathbf{c}) - f'(\mathbf{c})max(\mathbf{c})}{1 - f'(\mathbf{c})}, \tag{3.21}$$

from Equations 3.13 and 3.20, provided that $f'(\mathbf{c}) \neq 1$. However, if $w_{min} = 0$ and $f'(\mathbf{c}) = 1$, then $\ell(\mathbf{q}(\mathbf{c})) = M$ and thus we could not be in the case $\ell > \ell(\mathbf{q}(\mathbf{c}))$.

Finally, the (r, g, b) coordinates of $\mathbf{c}$ are the computed $max(\mathbf{c})$, $mid(\mathbf{c})$, and $min(\mathbf{c})$ values, where the order of $max(\mathbf{c})$, $mid(\mathbf{c})$, and $min(\mathbf{c})$ as r, g, and b are determined by the values of $k(\mathbf{c})$.

Algorithm *GLHS_TO_RGB*

Input: $w_{max}, w_{mid}, w_{min}$, such that $0 \leq w_{max}, w_{mid}, w_{min} \leq 1$, $w_{max} > 0$, and $w_{max} + w_{mid} + w_{min} = 1$.

$(\ell, h, s) \in R_A \cup R_B \cup R_C$, where
$R_A = [0, M] \times \{undefined\} \times \{0\}$,
$R_B = \emptyset$, if $w_{min} > 0$
$R_B = \{M\} \times \{60, 180, 300\} \times (0, 1]$, if $w_{min} = 0$ and $w_{mid} > 0$,
$R_B = \{M\} \times [0, 360) \times (0, 1]$, if $w_{min} = w_{mid} = 0$,
and $R_C = (0, M) \times [0, 360) \times (0, 1]$.

Output: $\mathbf{c} = (r, g, b)$, $r, g, b \in [0, M]$.

Auxiliary variables: the critical lightness $\ell(\mathbf{q})$, k, f, and f'.

```
begin
    if s = 0
        then r := g := b := ℓ;                      {achromatic}
        else begin                                  {chromatic}
            k := FLOOR(h/60);                       {sector-number}
            f := h/60 − k;                          {hue-within-the-sector}
            begin case of
                k even: f' := f;
                k odd: f' := 1 − f;
            endcase
            ℓ(q) := (w_mid * f' + w_max) * M;
            if ℓ ≤ ℓ(q)
                then begin
                    min := (1 − s) * ℓ;
                    mid := (f' * ℓ + min * ((1 − f') * w_max − f' * w_min))/
                           (w_max + f' * w_mid);
                    max := (ℓ − w_mid * mid − w_min * min)/w_max;
                end
                else begin                          {ℓ > ℓ(q)}
                    max := s * M + (1 − s) * ℓ;
                    mid := ((1 − f') * ℓ − max * ((1 − f') * w_max − f' * w_min))/
                           ((1 − f') * w_mid + w_min);
                    if w_min > 0
                        then min := (ℓ − w_max * max − w_mid * mid)/w_min;
                        else min := (mid − f' * max)/(1 − f');
                end
            case k of
                0: (r, g, b) := (max, mid, min);
                1: (r, g, b) := (mid, max, min);
                2: (r, g, b) := (min, max, mid);
                3: (r, g, b) := (min, mid, max);
                4: (r, g, b) := (mid, min, max);
                5: (r, g, b) := (max, min, mid);
            endcase
        end                                         {chromatic}
end; {GLHS_TO_RGB}
```

Figure 3.8 The GLHS-TO-RGB transformation algorithm.

3.5 ILLUSTRATIONS OF GLHS

We have implemented the GLHS model in an X-windows environment using the OSF/Motif toolkit. Two programs have been developed to support two applications: (i) color specification and coordinate transformations, described here, and (ii) coding of multiparameter distributions into integrated displays, which we discuss in Chapter 9.

3.5.1 Multiple Color Model Specification and Transformations System

MCMTRANS—Multiple Color Model Translation System lets the user specify a color as a vector in either RGB, GLHS, CIELUV [CIE78], or the Munsell Book of Color [Mun76]. The actual color selected is shown on a color patch, and its coordinates in all models are shown both numerically and on sliders. A "barycentric potentiometer" is provided to select the values of the three weights that select a specific model in GLHS. Figure 3.9 shows the interface of MCMTRANS and demonstrates the effect of the weights w_{max}, w_{mid}, and w_{min} on the appearance (and on the (r, g, b) coordinates) of the color described by the same LHS vector in different GLHS models.

The system provides a tool for the exploration and understanding of the relationships among color specifications in the various models.

3.5.2 Proof of the properties of the hue function

In this section we prove that the hue function, as defined by Equations 3.11–3.13, satisfies the properties specified in item 4 of Section 3.3. In order to simplify matters, we assume that $r > g \geq b$; i.e., that $k(\mathbf{c}) = 0$. The proofs for the other five sectors are similar. Under this assumption

$$f(\mathbf{c}) = \frac{g - b}{r - b}. \tag{3.22}$$

In Figure 3.10, we show the plane through the origin (black point), perpendicular to the main diagonal of the colorcube. All coordinates in the figure

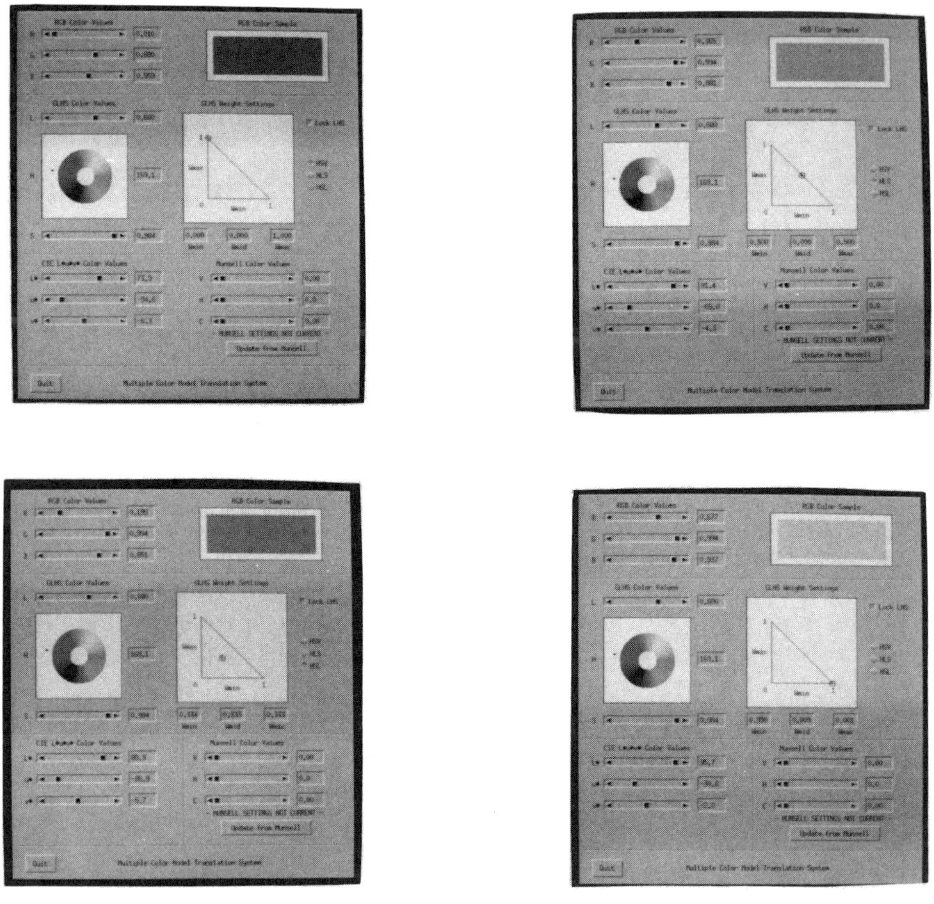

Figure 3.9 The MCMTRANS color specification program. A fixed LHS vector ($\ell = 0.68, h = 169.1, s = 0.984$) is converted using four GLHS models (defined by the weights in the barycentric potentiometer). Note the difference in appearance and in RGB coordinates resulting from the different models. Top left: $(w_{max}, w_{mid}, w_{min}) = (1,0,0)$ (HSV-hexcone); Top right: $(w_{max}, w_{mid}, w_{min}) = (0.5,0,0.5)$ (HLS-double-hexcone); Bottom left: $(w_{max}, w_{mid}, w_{min}) = (0.333,0.333,0.334)$ (LHS-triangle); Bottom right: $(w_{max}, w_{mid}, w_{min}) = (0.001,0.009,0.99)$. (See Web site for color version.)

are in the (r, g, b)-space. The point $\mathbf{p}(\mathbf{c})$ is the projected color vector of $\mathbf{c}$. Its coordinates are $\left(\frac{2r-g-b}{3}, \frac{2g-b-r}{3}, \frac{2b-r-g}{3}\right)$; see item 3 in Section 3.3.

Consider now the set of colors $\mathbf{c}' = (r', g', b')$ which satisfy the linear equation

$$f(\mathbf{c})r' - g' + (1 - f(\mathbf{c}))b' = 0. \tag{3.23}$$

Clearly, this set forms a planar segment of the colorcube, which contains $\mathbf{c}$ and all achromatic colors. For all chromatic colors $\mathbf{c}'$ satisfying Equation 3.23 for which $k(\mathbf{c}') = 0$, $f(\mathbf{c}') = f(\mathbf{c})$ and the hue of $\mathbf{c}'$ is the same as the hue of $\mathbf{c}$.

Consider now the pure red $(t, 0, 0)$, where $t = 1/\sqrt{2}$. Its projected color vector is $\mathbf{r} = (2t/3, -t/3, -t/3)$. Let θ be the angle between $\mathbf{r}$ and $\mathbf{p}(\mathbf{c})$. Consider also the pure green $(0, t, 0)$. Its projected color vector is $\mathbf{g} = (-t/3, 2t/3, -t/3)$. The distance between $\mathbf{r}$ and $\mathbf{g}$ is 1. Let $\mathbf{p}(\theta)$ be the vector which is at the intersection of the line containing $\mathbf{r}$ and $\mathbf{g}$ and the line containing the origin and $\mathbf{p}(\mathbf{c})$. Let $x(\theta)$ denote the distance from $\mathbf{r}$ to $\mathbf{p}(\theta)$. We note that for chromatic colors $\mathbf{c}$ such that $k(\mathbf{c}) = 0$, $0 \leq x(\theta) < 1/2$. Clearly, $\mathbf{p}(\theta)$ is the projected color vector of the color $(t(1 - x(\theta)), tx(\theta), 0)$, which has the same hue as $\mathbf{c}$ (since it lies on the planar segment of the colorcube which contains $\mathbf{c}$ and the achromatic axis). Substituting the components of this vector into Equation 3.22 we get that

$$f(\mathbf{c}) = \frac{x(\theta)}{1 - x(\theta)}. \tag{3.24}$$

This shows that the hue of a chromatic color $\mathbf{c}$ is indeed a function of the angle between its projected color vector $\mathbf{p}(\mathbf{c})$ and the projected color vector $\mathbf{r}$ of a pure red. The domain of θ for $k(\mathbf{c}) = 0$ is $[0, 60)$, the corresponding values of $f(\mathbf{c})$ range over $[0, 1)$, and so, by Equation 3.11, $h(\mathbf{c})$ ranges over $[0, 60)$. By similar arguments for the other five sectors, we see that 4a of Section 3.3 is satisfied. It is also trivial to see from Equations 3.11, 3.12, 3.24 and Figure 3.10 that hue is a continuous and monotonically increasing function of θ, as stated in 4b of Section 3.3.

We are left to show 4c of Section 3.3. To do this, it is easier to use radians rather than degrees for angles. The claim is then that the absolute value of the difference function

$$d(\theta) = \frac{x(\theta)}{1 - x(\theta)} \cdot \frac{\pi}{3} - \theta \tag{3.25}$$

is "small" for $0 \leq \theta < \pi/3$.

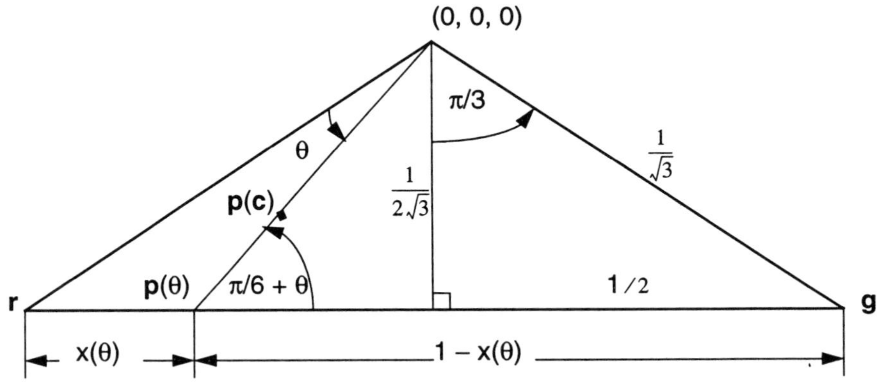

Figure 3.10 The plane through the origin perpendicular to the main diagonal of the colorcube.

Using the fact that (see Figure 3.10)

$$\tan(\frac{\pi}{6} + \theta) = \frac{1}{2\sqrt{3}(1/2 - x(\theta))}, \tag{3.26}$$

we get the expression that

$$d(\theta) = \frac{\sqrt{3}\tan(\frac{\pi}{6} + \theta) - 1}{\sqrt{3}\tan(\frac{\pi}{6} + \theta) + 1} \cdot \frac{\pi}{3} - \theta. \tag{3.27}$$

We note that

$$d(0) = 0 \text{ and } \lim_{\theta \to \pi/3} d(\theta) = 0. \tag{3.28}$$

Since $d(\theta)$ is differentiable in $[0, \pi/3)$, it acquires its maxima and minima in that range at points which are easily found by having the property $d'(\theta) = 0$. Straightforward manipulation shows that this condition is satisfied if

$$\left[\sqrt{3}\sin(\frac{\pi}{6} + \theta) + \cos(\frac{\pi}{6} + \theta)\right]^2 = \frac{2\pi}{\sqrt{3}}. \tag{3.29}$$

Trivial trigonometrical rearrangement shows this to be equivalent to

$$\left[\cos(\theta - \frac{\pi}{6})\right]^2 = \frac{\pi}{2\sqrt{3}}. \tag{3.30}$$

The two angles in $[0, \pi/3)$ which satisfy Equation 3.30 are (to three significant digits) 0.213 and 0.833. Substituting theses values into Equation 3.27 shows that the absolute value of the difference function is always less than 0.0195 radians, and so less than 1.12 degrees, as claimed in Section 2.1

3.6 SUMMARY AND NOTES

We have discussed color models for computer graphics. The focus was on the Lightness, Hue, and Saturation family of models. We have shown that these can be described in a unifying framework. That framework has been presented in the form of the Generalized Lightness, Hue, and Saturation (GLHS) model. We have shown an implementation of the GLHS model. Its potential use as a coding mechanism to generate integrated displays of multiparameter distributions is illustrated in Chapter 9.

The GLHS family provides a clean unified way of describing LHS models and interacting with them. The algorithms presented translate to and from RGB for any of HSL, HSV, and HLS (as well as any other special case of the generalized model) and yet they are as simple as the separate algorithms given in books for translation to and from RGB for the specific color models. The GLHS family also introduces new LHS models that have not been explored previously. Such models may be advantageous to achieve better properties within an LHS model, such as uniformity [LH88, LX92]. In addition, the GLHS model provides the potential for dynamic color model changes via interaction with the weights, given sufficient computational power, see Chapter 10. Such capabilities may enhance the detection of features in color displays.

On the negative side, the known models within GLHS suffer from the same drawbacks as many other mathematical models of color. In particular, most mathematical definitions of lightness fall short of adequately describing the process of lightness perception in human color vision. Thus, colors that are (mathematically) classified as having the same lightness, often do not appear as such. Similarly, the mathematical modeling of hue and saturation do not adequately represent their perception by humans. Thus, these geometrical models of color (sometimes referred to as pseudo perceptual) are a simplification of color vision, and are only perceptual in quality (they use the perceptual primaries lightness, hue, and saturation for their axes) and not in quantity (the scaling and calibration of each axis). The CIELUV and CIELAB models are claimed to be perceptually proven, though there seem to be certain limitations to that claim; in particular, their uniformity is only local (see Sections 2.7 and 2.7). Even so, they have been advocated as adequate representations of human color perception and as appropriate coding mechanisms for multiparameter image data [Rob85, RO85, RO86, Taj83a, Taj83b].

One promise that the GLHS family presents is the potential for finding within the family a model that shares the perceptual properties of a proven uniform model and, at the same time, the algorithmic properties of the GLHS family. Finding such a GLHS model requires a search among all GLHS models for one that is the closest approximation to the selected uniform model, subject to some predefined criteria for closeness. We describe the details of such efforts in Chapter 4.

4

MULHS—A MOST UNIFORM LHS MODEL: APPROXIMATING UNIFORM COLOR SPACES WITH THE GLHS FAMILY OF MODELS

In many applications, e.g., computer graphics representation of parameter distributions (which we discuss in detail in Chapters 7 and 9) several properties of the color mapping are desirable in order to make effective use of color. *Uniformity*—the property that maps values to colors whose perceptual distances correspond to the distances between the values—is one of them.

We recall that a *Uniform Color Space (UCS)* is a color model in which distances between points adequately represent perceptual distances between the colors represented by these points. One of the recommended used UCSs for color monitors is the CIELUV uniform color space, see Chapter 2 and [CIE78].

GLHS—the generalized lightness, hue, and saturation color model—enables the realization of different models as special cases by specifying its weight values.

In this chapter, we explore the search for a member of the GLHS family that adequately approximates a uniform color space. Such a space would provide both uniformity and the algorithmic properties of the GLHS family, thus facilitating the use of a uniform space in computer applications, [LH88, LX92].

We discuss the search for a particular set of GLHS weights that will yield a specific GLHS model that approximates a given uniform space. This introduces an optimization problem, where we try to minimize an objective function that represents the difference between the given uniform space and a GLHS space that is expected to approximate it.

We first look at the general problem. We then look at two specific instances: an approximation to the CIELUV space, followed by an approximation to the Munsell Book of Color.

4.1 INTRODUCTION: A MINIMIZATION PROBLEM

The GLHS models provide transformation algorithms between the GLHS and RGB spaces. Given the transformations between any uniform color space U and the RGB space, the composition of transformations defines a transformation between the GLHS space and U.

Let (u_1, u_2, u_3) denote the axes of U. For any two colors $\mathbf{c}, \mathbf{c}'$, their distance $d_U(\mathbf{c}, \mathbf{c}')$ in U space is defined as the Euclidean distance

$$d(\mathbf{c}, \mathbf{c}') = \sqrt{(u_1(\mathbf{c}) - u_1(\mathbf{c}'))^2 + (u_2(\mathbf{c}) - u_2(\mathbf{c}'))^2 + (u_3(\mathbf{c}) - u_3(\mathbf{c}'))^2}. \quad (4.1)$$

For any three colors $\mathbf{c}, \mathbf{c}', \mathbf{c}''$, the (second order) distance $d'(\mathbf{c}, \mathbf{c}', \mathbf{c}'')$ between their distances is defined as

$$d'(\mathbf{c}, \mathbf{c}', \mathbf{c}'') = |d(\mathbf{c}, \mathbf{c}') - d(\mathbf{c}', \mathbf{c}'')|. \quad (4.2)$$

The uniformity of U means that for any three colors $\mathbf{c}, \mathbf{c}', \mathbf{c}''$ resulting from regular increments along any of the axes (u_1, u_2, u_3), their second order distance d' will be zero (since $d(\mathbf{c}, \mathbf{c}') = d(\mathbf{c}', \mathbf{c}'')$). For example, let $\mathbf{c} = (u_1, u_2, u_3)$, $\mathbf{c}' = (u_1 + \Delta u_1, u_2, u_3)$, $\mathbf{c}'' = (u_1 + 2\Delta u_1, u_2, u_3)$. Then, $d(\mathbf{c}, \mathbf{c}') = d(\mathbf{c}', \mathbf{c}'') = \Delta u_1$ and $d'(\mathbf{c}, \mathbf{c}', \mathbf{c}'') = 0$.

Let $T : GLHS \to U$ be the transformation from GLHS to U. For any weight assignment in GLHS and for any color $\mathbf{c} = (l, h, s)$ in the space G defined by the particular weights chosen, $T(\mathbf{c}) = (u_1(\mathbf{c}), u_2(\mathbf{c}), u_3(\mathbf{c}))$. If G is uniform, then the Laplacian

$$\nabla^2 T(l, h, s) = \frac{\partial^2 T}{\partial l^2}(l, h, s) + \frac{\partial^2 T}{\partial h^2}(l, h, s) + \frac{\partial^2 T}{\partial s^2}(l, h, s), \quad (4.3)$$

of T should be equal to zero for all points (l, h, s).

If there does not exist a set of weights such that the resulting space is uniform, the space that would be the closest approximation to U would be the space for which $\int_{(l,h,s) \in GLHS} |\nabla^2 T(l, h, s)| dl dh ds$ is minimum.

Thus the problem of searching the best approximation in the GLHS family of models can be stated as:

Find the space G in GLHS (defined by the set of weights $w_{max}, w_{mid}, w_{min}$) that minimizes

$$\int_{(l,h,s)\in GLHS} |\nabla^2 T(l,h,s)| dl dh ds. \tag{4.4}$$

Instead of solving the continuous problem, we look for the set of weights $w_{max}, w_{mid}, w_{min}$ that minimizes

$$\sum_{(l,h,s)\in GLHS} \left(\begin{array}{cc} & \frac{d'((l,h,s),(l+\Delta l,h,s),(l+2\Delta l,h,s))}{(\Delta l)^2} \\ + & \frac{d'((l,h,s),(l,h+\Delta h,s),(l,h+2\Delta h,s))}{(\Delta h)^2} \\ + & \frac{d'((l,h,s),(l,h,s+\Delta s),(l,h,s+2\Delta s))}{(\Delta s)^2} \end{array} \right). \tag{4.5}$$

4.2 GLHS APPROXIMATION OF THE CIELUV UNIFORM COLOR SPACE

The CIE recommendation that defined the CIELUV Uniform Color Space specifies the transformation formulae between the RGB and CIELUV spaces. The generalized LHS model GLHS provides us with transformation algorithms between the GLHS and RGB spaces. Their composition defines a transformation between the GLHS and CIELUV spaces.

Referring to the formulation in Section 4.1, let $U = CIELUV$. Thus, $u_1 = L^*, u_2 = u^*, u_3 = v^*$. Then, for any two colors $\mathbf{c}, \mathbf{c}'$, Equation 4.1 becomes

$$d(\mathbf{c}, \mathbf{c}') = \sqrt{(L^*(\mathbf{c}) - L^*(\mathbf{c}'))^2 + (u^*(\mathbf{c}) - u^*(\mathbf{c}'))^2 + (v^*(\mathbf{c}) - v^*(\mathbf{c}'))^2}. \tag{4.6}$$

As stated previously for the general UCS U, the uniformity of the CIELUV space means that for any three colors resulting from regular increments along one of the axes (L^*, u^*, v^*) the distance d' between their distances will be zero. For example, let $\mathbf{c} = (L^*, u^*, v^*), \mathbf{c}' = (L^* + \Delta L^*, u^*, v^*), \mathbf{c}'' = (L^* + 2\Delta L^*, u^*, v^*)$. Then, $d(\mathbf{c}, \mathbf{c}') = d(\mathbf{c}', \mathbf{c}'') = \Delta L^*$ and $d'(\mathbf{c}, \mathbf{c}', \mathbf{c}'') = 0$.

Let $T : GLHS \rightarrow CIELUV$ be the transformation from GLHS to CIELUV. The remainder of the optimization problem is stated precisely as it was stated for the general case in Equations 4.3–4.5 in Section 4.1.

4.2.1 Results

As Levkowitz and Herman reported in [LH88], we have computed the expression of Equation 4.5 for various choices of the weights $w_{max}, w_{mid}, w_{min}$. None of the choices yielded a sum of zero (that is, none of the spaces defined by our choices is uniform according to the uniformity criterion that we have chosen). However, the sum has the lowest values for weight assignments for which the value of w_{max} is close to the value of w_{min}. Such models are close to the triangular models ($w_{max} = w_{mid} = w_{min} = 1/3$) and to the HSL-Double-Hexcone Model ($w_{max} = w_{min} = 1/2, w_{mid} = 0$). This result supports the claim that the triangular models provide a reasonable approximation (among the available LHS models) to a uniform color space [Far83]. More specifically, the GLHS generalized family of models provides us with an even closer approximation by choosing $w_{max} = 0.5, w_{mid} = 0.02, w_{min} = 0.48$. See [LH88, Lev88] for details.

4.3 GLHS APPROXIMATION OF THE MUNSELL BOOK OF COLOR

Following our general discussion on the approximation of a uniform color space with a GLHS model, and our specific discussion of such approximation of the CIELUV uniform color space, we now describe a similar approximation of another uniform space, the Munsell Book of Color, see Section 2.7, [Mun05, Mun76].

Our Munsell Book of Color space is a polar system where $0 \leq C \leq 15$, $0 \leq H < 40$, $0 \leq V \leq 10$. We define Cartesian coordinates (x_m, y_m, z_m) in Munsell space as $x_m = 10C\cos(9H)$, $y_m = 10C\sin(9H)$, and $z_m = 10V$. (The multiplication by 9 maps the hue range from $[0, 40)$ to $[0, 360)$; the multiplication by 10 maps the chroma range from $[0, 15]$ to $[0, 150]$, and the value range from $[0, 10]$ to $[0, 100]$. The purpose of these range mappings is to make the ranges of the Munsell and CIELUV spaces similar.)

Then the color difference in the Munsell Book space is defined (by applying the Munsell axes to Equation 4.1) as

$$d_{Munsell}(\mathbf{c_1}, \mathbf{c_2}) = \sqrt{\Delta x_m^2 + \Delta y_m^2 + \Delta z_m^2}. \tag{4.7}$$

Our coordinate transformation is $(l, h, s) \mapsto (r, g, b) \mapsto (X, Y, Z) \mapsto (x, y, Y) \mapsto (H, V, C)$, where $H_i \leq H \leq H_{i+1}$, $V_j \leq V \leq V_{j+1}$, and $C_k \leq C \leq C_{k+1}$,

Model	Weights w_{max}	w_{mid}	w_{min}	Average difference
Hexcone	1	0	0	0.279776
Double hexcone	0.5	0	0.5	0.310995
Triangle	0.34	0.33	0.33	0.329203
Minimizer	0.7	0.1	0.2	0.252529

Table 4.1 Average difference values for the minimizer and the hexcone and double hexcone models.

and $H_i, H_{i+1}, V_j, V_{j+1}, C_k, C_{k+1}$ are integer values within the Munsell Book data. Finally, we locate the nearest neighbor in the Munsell data of the point (H, V, C) using a triple binary search. This defines the transformation $T : GLHS \rightarrow Munsell$.

4.3.1 Results and comparison

As reported by Levkowitz and Xu in [LX92], we have computed the expression in Equation 4.5 in the Munsell space with all possible values of the weights w_{max},w_{mid},w_{min} at 0.1 increments. Table 4.1 shows the average differences for the hexcone (weights (1, 0, 0)), double hexcone (weights (0.5, 0, 0.5)), and triangle (weights (0.34, 0.33, 0.33)) models, and the minimizer model (weights (0.7, 0.1, 0.2). All computations were made with $\Delta l = \Delta s = 0.1$ and $\Delta h = 1^o$ around points sampled at 0.2 intervals along l and s, and 10 degree intervals along h. Note that since the Munsell space is smaller than the GLHS spaces, the results were computed only using points in the GLHS space that map to values inside the Munsell space. (In [Xu96], Xu extended the Munsell space.)

While no computed points yielded a sum of zero, (that is, none of the models defined by our choices of weights is identical to the Munsell space), the space defined by the weight values $w_{max} = 0.7$, $w_{mid} = 0.1$, and $w_{min} = 0.2$ represents the closest approximation to the Munsell space under the current optimization process.

Figure 4.1 shows the loci in (u^*, v^*)-coordinates of constant GLHS hue (straight lines radiating from the center) and saturation (concentric hexagons) at a fixed lightness ($l = 0.5$) for the model that minimizes the difference (weights (0.7, 0.1, 0.2)). Note that the straight lines of constant GLHS hue remain straight after the transformation to (u^*, v^*)-coordinates. This indicates some match between the two spaces [Hun87]. However, the match is not perfect, as indicated by the hexagonal constant saturation curves (a perfect match would yield circles). Figure 4.2 shows the same plots for the hexcone model (weights (1, 0, 0)) for comparison. Note the similarities between the two.

Figure 4.3 shows a GLHS hue plane ($h = 0$) plotted in (L^*, C^*_{uv})-coordinates for the model that minimizes the difference (weights (0.7, 0.1, 0.2)). Here, for a perfect match, neighboring points should form squares [Hun87], which is clearly not the case. Again, for comparison, Figure 4.4 shows the same plots for the hexcone model.

Figure 4.5 shows the loci in (u^*, v^*)-coordinates of constant Munsell hue and chroma at Munsell value of 5. Figure 4.6 shows a Munsell hue plane ($H = 0$) in (L^*, C^*_{uv})-coordinates. The plots indicate a relatively close match between the Munsell and the CIELUV spaces for $u^* \geq 0$ and $v^* \leq 0$ (second quadrant in Figure 4.5), where the curves are almost perfect circles, and for $L^* > 30$ (in Figure 4.6), where constant value and chroma lines form squares. However, there is no close match in the other regions—note the constant hue curvatures, the compressed chroma curves in the fourth quadrant, and the protruding curves in the first and third quadrants in Figure 4.5; and the slant rectangles for $L^* < 30$ in Figure 4.6.

4.4 SUMMARY AND NOTES

We have discussed the importance of uniformity of the coordinates of color models. We have described the potential role of GLHS as a framework for developing a color model that has these desired properties—a uniform LHS color model. We have stated a minimization problem for selecting the GLHS model that is the closest approximation to a given uniform color space in general. We have further stated the particular minimization problems necessary to find closest approximations to the Munsell Book of Color and CIELUV spaces, respectively.

The results of such minimization computations show that while none of the LHS models used in these computations is uniform, the choice of weights does affect the closeness of a model to uniformity.

Other potential candidates for such approximation may be Xu's extended Munsell space [Xu96]; the TekHVC model, which is based on CIELUV [TMM89]; the Optical Society of America (OSA) Uniform Color Scale [MG80, MG87];

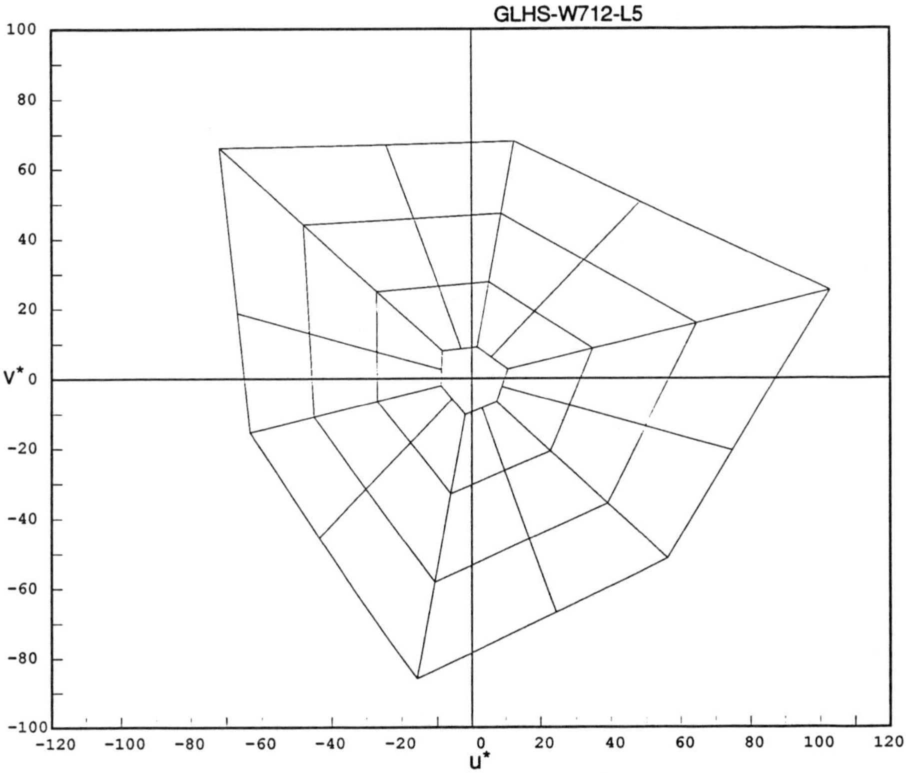

Figure 4.1 Constant GLHS hue and saturation curves in (u^*, v^*)-coordinates: For the minimizer space ($w_{max} = 0.7$, $w_{mid} = 0.1$, $w_{min} = 0.2$). From [LX92].

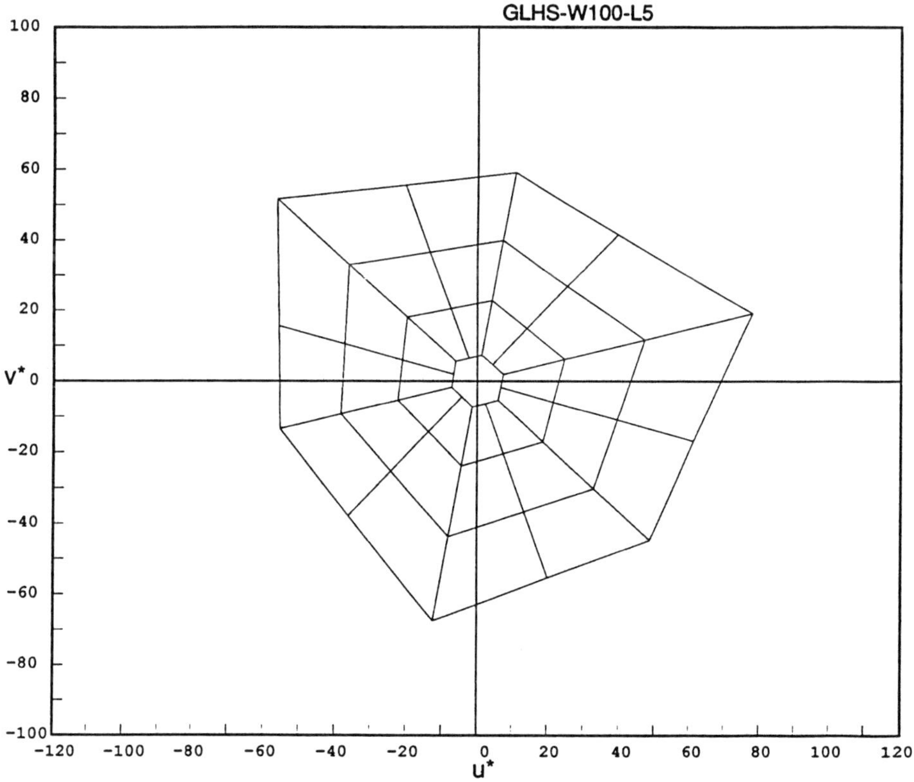

Figure 4.2 Constant GLHS hue and saturation curves in (u^*, v^*)-coordinates: For the hexcone space ($w_{max} = 1$, $w_{mid} = w_{min} = 0$.) From [LX92].

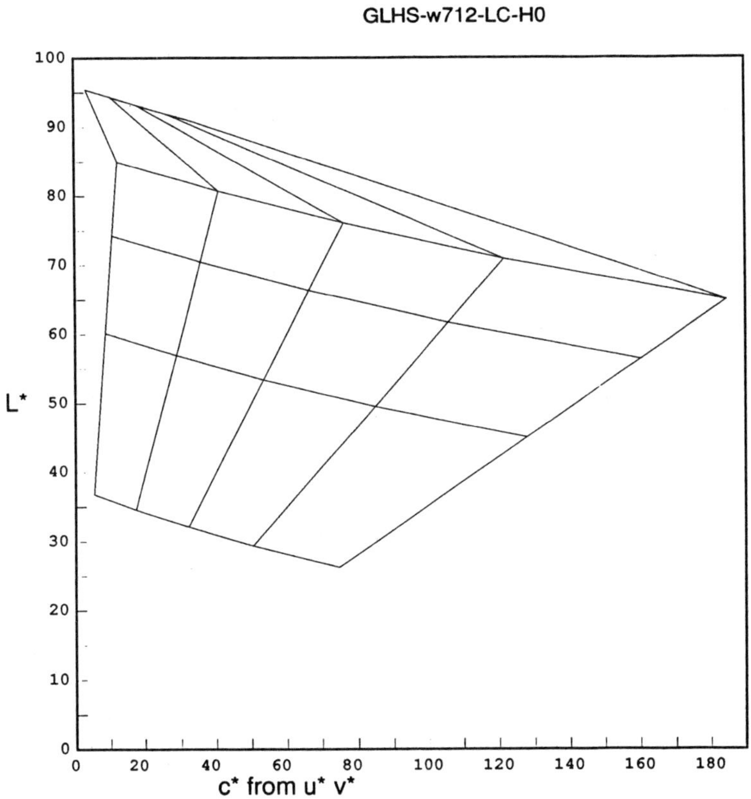

Figure 4.3 A GLHS hue plane ($h = 0$) in (L^*, C^*_{uv})-coordinates: For the minimizer space ($w_{max} = 0.7$, $w_{mid} = 0.1$, $w_{min} = 0.2$). From [LX92].

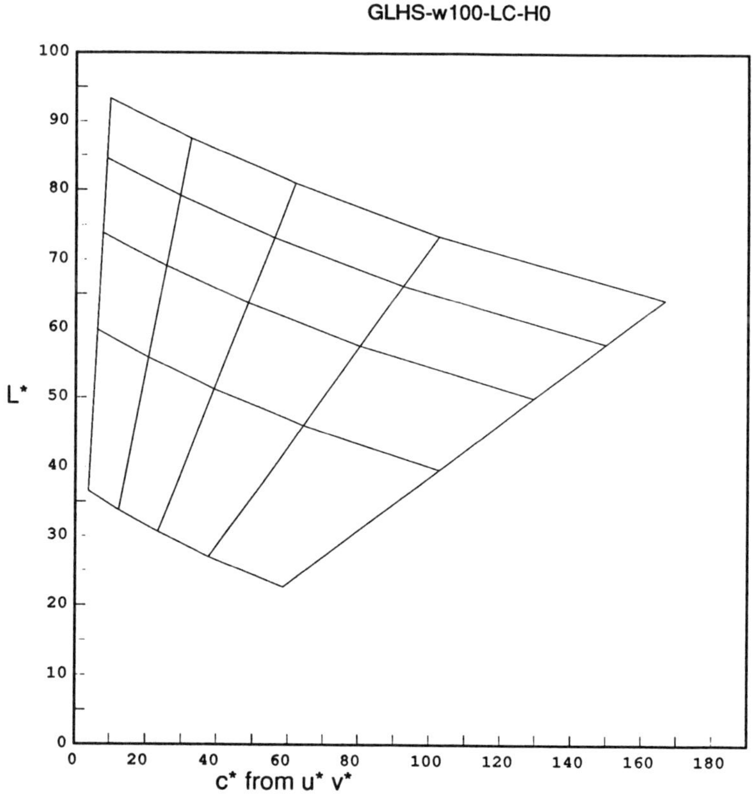

Figure 4.4 A GLHS hue plane ($h = 0$) in (L^*, C^*_{uv})-coordinates: For the hexcone space ($w_{max} = 1$, $w_{mid} = w_{min} = 0$.) From [LX92].

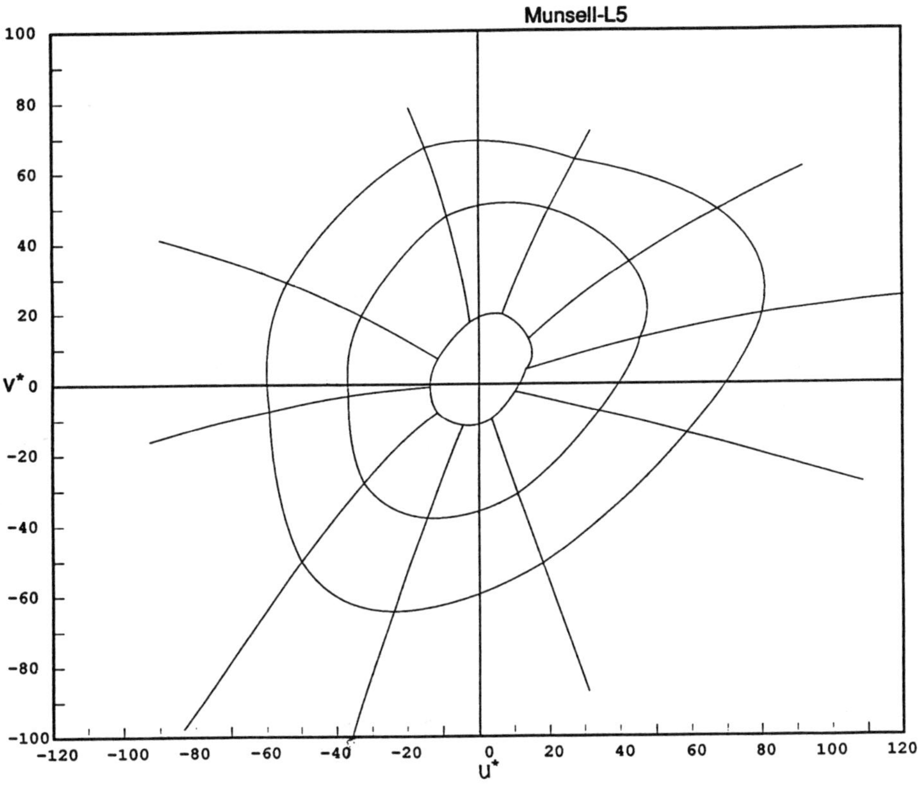

Figure 4.5 Constant Munsell hue and chroma curves in (u^*, v^*)-coordinates, $(L = 5)$. From [LX92].

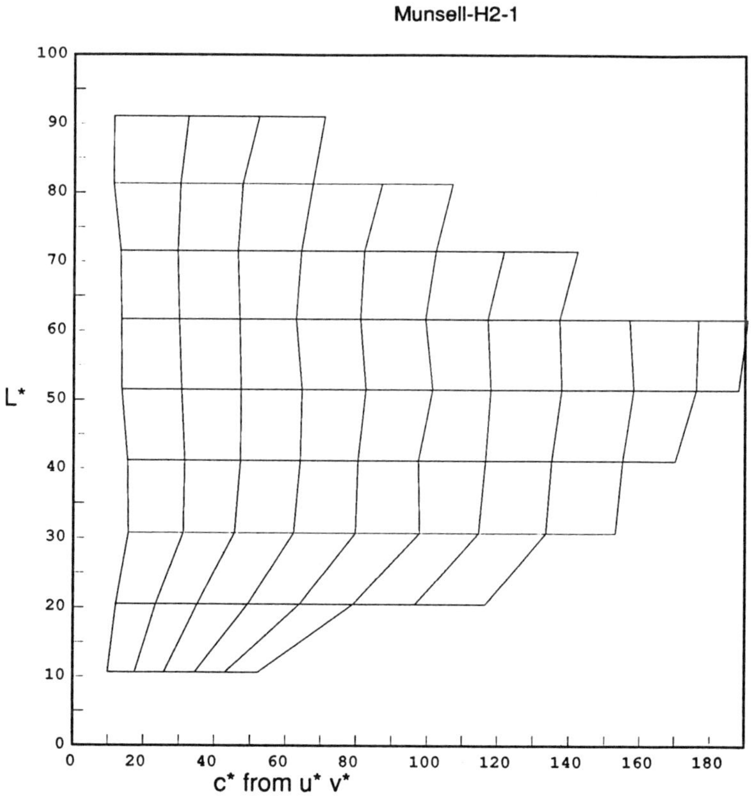

Figure 4.6 Constant Munsell hue and chroma curves in (u^*, v^*)-coordinates, $(L = 5)$: A Munsell hue plane $(H = 0)$ in (L^*, C^*_{uv})-coordinates. From [LX92].

and models based on Luo and Rigg's color-differences formula [LR86, LR87a, LR87b, LR87c].

The search for GLHS approximations to uniform color spaces will come to a successful conclusion only when psychophysical studies show that mathematically optimal GLHS approximations of uniform spaces share the desirable perceptual properties of the uniform color spaces that they were intended to approximate.

The question whether uniformity offers a significant advantage in the use of a color model in various applications is a matter for further studies.

5

COLOR VISION FOR COMPLEX VISUAL TASKS

Color plays an important role in almost all visual tasks. Without thinking much about it, we are continuously engaged in decision processes that are all driven by color information input. Crossing an intersection, selecting a fruit or vegetable at the market, deciding whether the bread in the oven is ready, are just a small sample of all those tasks that are primarily decided upon based on color analysis.

Color can be a great facilitator in search tasks, though, if not used properly, it can also hinder the task. And that is true of other tasks as well.

We examine color's influence on a number of tasks now.

5.1 COLOR AND VISUAL SEARCH

Color attracts "preattentive" vision. In other words, it attracts attention, facilitates effortless search, and enables parallel search. That is, search that is independent of the number of items being searched, and that is accomplished relentlessly, without interfering with any cognitive tasks being performed at the same time.

This has many applications in day-to-day activities. Identifying your yellow-with-Rustoleum-spots car in a packed mall parking lot, or your colorfully-dressed child inside that mall just before Christmas. Drawing attention to features of interest, as can be seen all that commonly in shopping areas. Scanning for multiple instances of a feature (for example, a searched word in a Web browser), and many others.

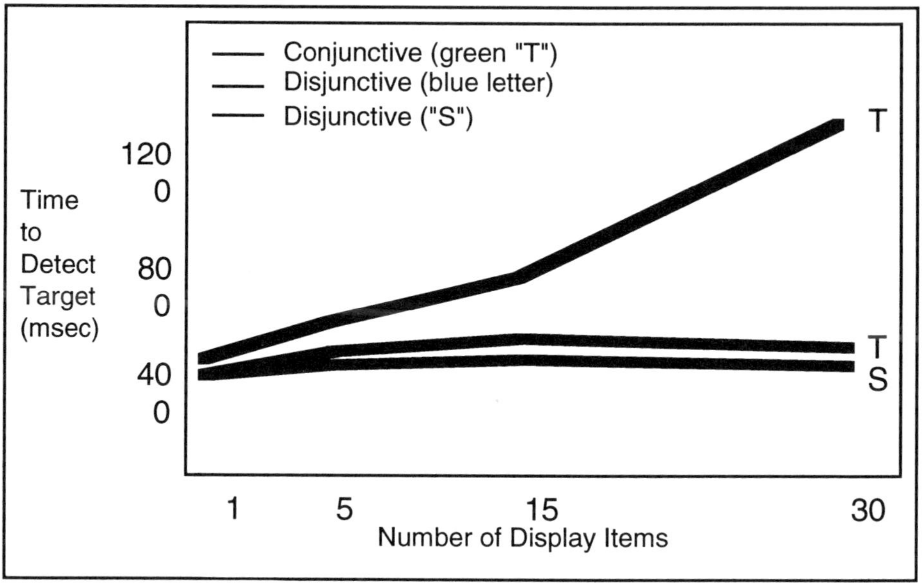

Figure 5.1 Preattentive search: Treisman and Gelade (1980).

The theoretic basis is Treisman's research on the "Popout" effect in visual search [TG80]. What Treisman found out in that research was that the time required to detect the presence of a target will be independent of the number of distracting items—in other words, the search will be parallel—only if the target is disjunctive, that is, based on a single feature 9e.g., angle, length, shape, color). Targets based on a conjunction of features require search times that increase linearly with the number of distracting items—i.e., serial search, see Figure 5.1.

Nakayama found out that color and depth are the most salient "popout" features, [NS86]. In other words, targets that are distinguished from their distractors by color or by depth will be more detectable than objects that are not, even if they are distinguishable by other features.

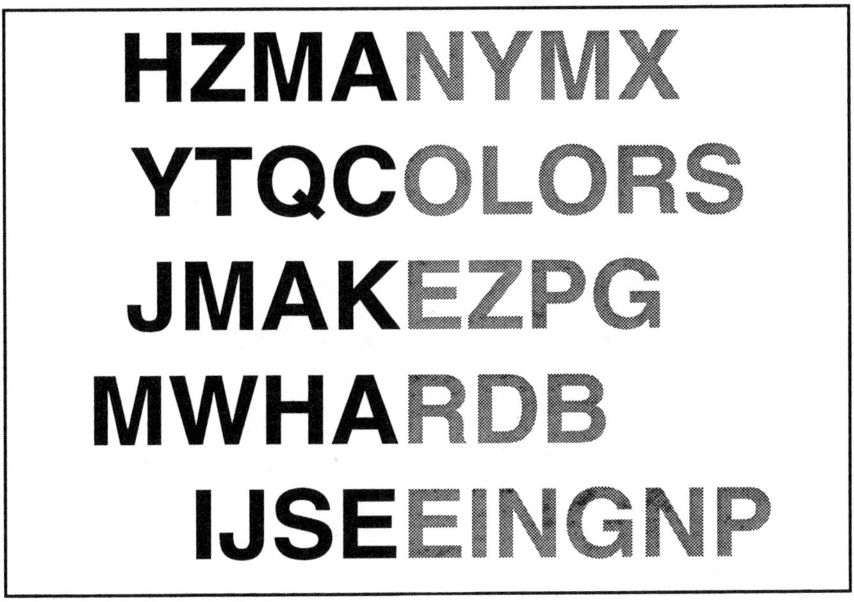

Figure 5.2 Preattentive color features distract. (See Web site for color version.)

However, a few caveats regarding using color for feature coding should be observed:

1. Preattentive features can distract the user from more effortful functions. In Figure 5.2, the color assignment distracts the viewer from the message. However, a better color assignment, as seen in Figure 5.3 can help.

2. Color coding can unwittingly engage the powerful preattentive aspects of human vision for trivial ends.

3. The role of preattentive color search can be overstated—processes requiring attention, scrutiny, and analysis are also important.

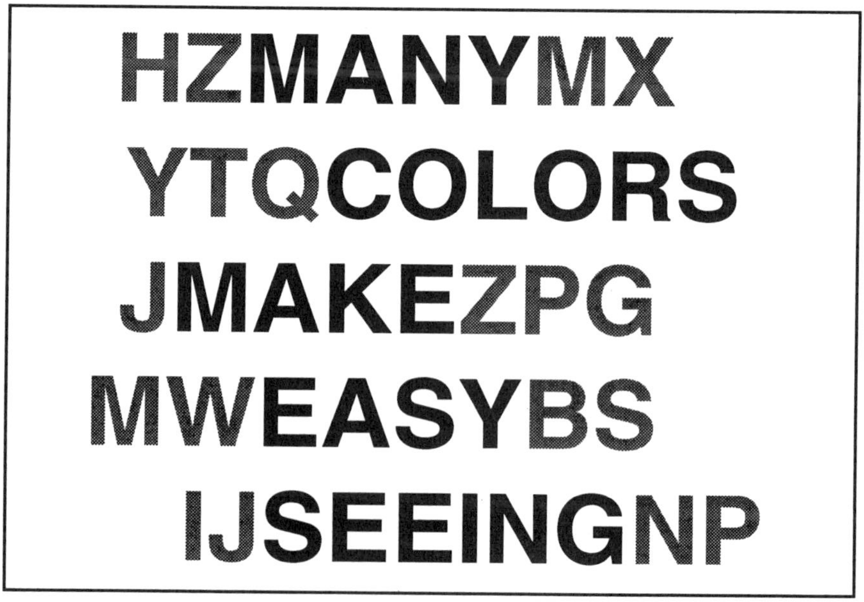

Figure 5.3 Preattentive color features help. (See Web site for color version).

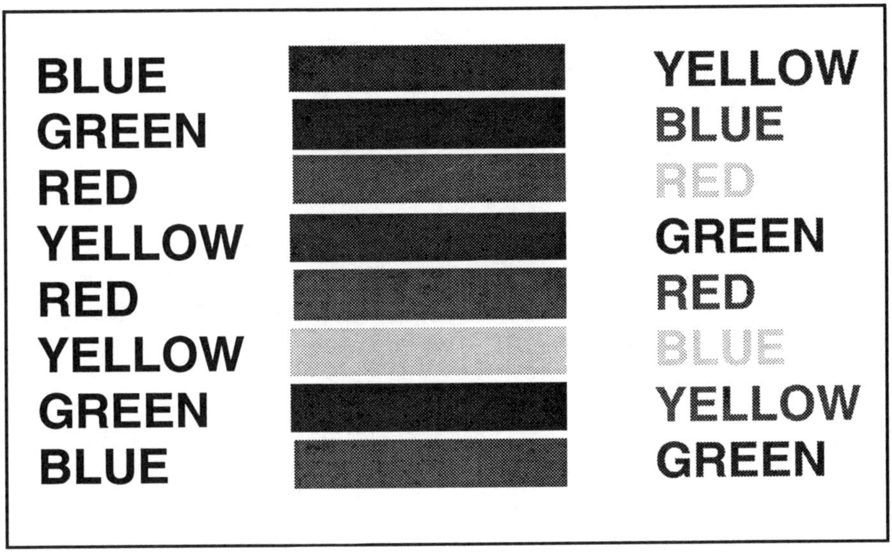

Figure 5.4 Visual-verbal interaction, after [Sch90a]. (See Web site for a color version.)

5.2 VISUAL-VERBAL INTERACTIONS: THE STROOP EFFECT (1935)

One interesting phenomenon is the power color perception has over verbal capabilities. Consider the following task: name the ink color of various words under two different conditions.

1. If the words are nonsense words—color naming is very fast.
2. If the words are color names, and each describes a different color than the color in which they are written, color naming is very slow, See Figure 5.4.

The reason is that reading a meaningful word operates quickly, preattentively, thus inhibiting the response to the color naming task [Str35].

This illustrates the importance of considering the interactions among the various channels that are being activated when putting a display. See Figure 5.4.

5.3 COLOR DISCRIMINATION VS. COLOR NAMING

Color discrimination, the side-by-side comparison of two color patches is extremely powerful, and is very comparable in all color-normal viewers. It can be described by a W-shaped function of wavelength (λ), $\Delta\lambda$. For very short, mid-spectral, and very long wavelengths, color discrimination is limited to 5–10 nm differences in wavelength; for the interstitial values, the ability to discriminate colors is five to ten times better, at 1 nm difference.

Color Naming, on the other hand, is highly individual, depending on training and culture, among other factors. For example, fabric designers of both sexes, as well as women, irrespective of their occupation, seem to know and use more color names, even though they have no better color discrimination than the rest of the population.

However, color naming does not interfere with color discrimination; even cultures that have only two color names (dark and light) perform normally on color discrimination tasks.

It has been shown that hues group perceptually into discrete categories, with the most salient categories having simple color names. There are eleven colors that are almost never confused [Boy89]. And color memory is best for those eleven fundamental colors.

Color naming can be used to characterize color appearance [AGC90].

5.4 COLOR CONTRAST AND COLOR CONSTANCY

Color is colors, plural. —Josef Albers [Alb75]

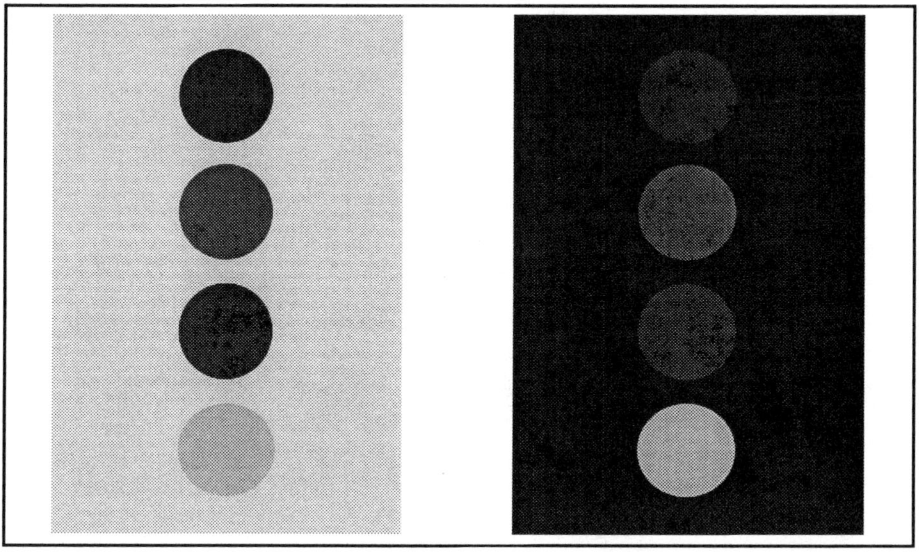

Figure 5.5 Perceived color depends on the color of background. Each pair of dots has the same chromaticity. (See Web site for a color version.)

The perception of color in general, and of color contrast in particular, depends on color context. That is, perceived color (hue) depends on the surrounding hues. There is an induction of complementary hues of the perceived-hue-context surrounding colors, which is mediated by the opponent process mechanisms, see Figures Figure 5.5 and 5.6.

It is, thus, difficult to analyze colors in a complex color scene, and to make color assessments without taking into consideration the surrounding colors. That, as we see later, has been one of the most important shortcomings of most color coding schemes in imaging.

Color constancy is the ability to discern the correct hue of an object, and to perceive hues as relatively constant, independent of the illumination. Consider observing a red object under white illumination, and then the same object, but

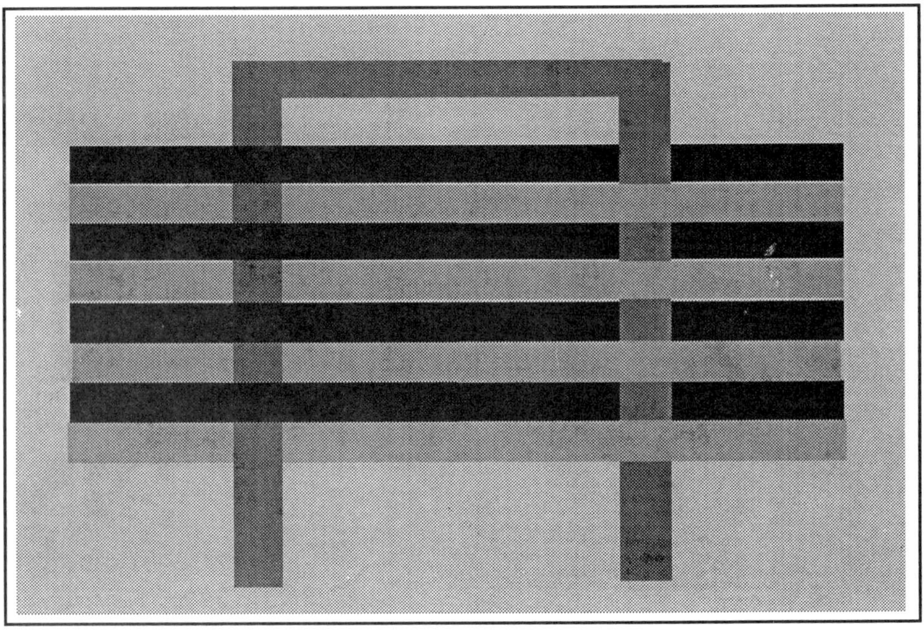

Figure 5.6 Induced color. The gray arch looks yellow or blue, depending on its surround (after a painting by Ken Knoblauch). (See Web site for a color version.)

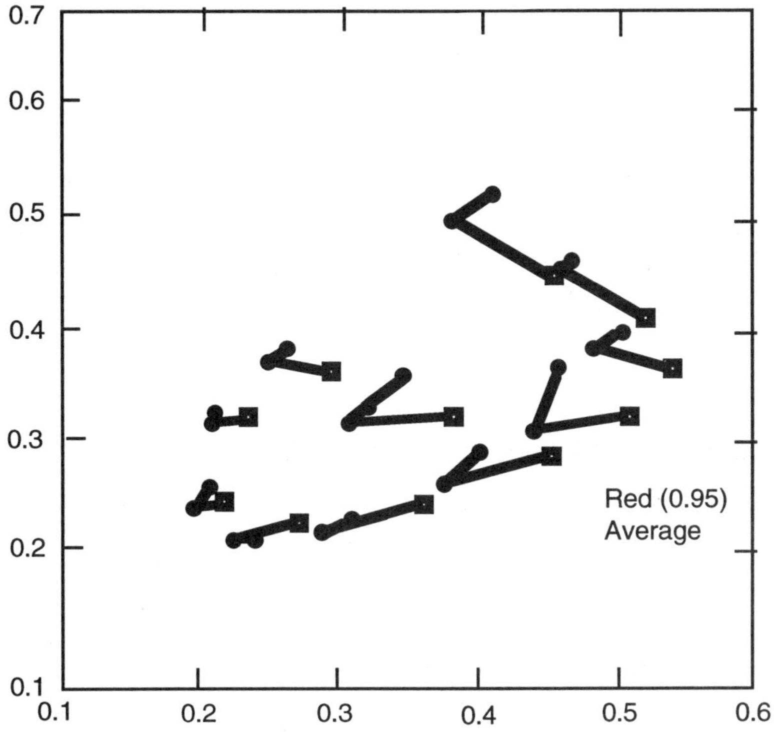

Figure 5.7 Color Constancy isn't Perfect (after [WBR87]).

white, under red illumination. Under most circumstances, human observers, after a short period of adaptation to the change in illumination, will be able to tell the correct hue of either object.

This involves a process called *chromatic adaptation* [Xu96], where the ambient illumination is "discounted" from the scene [von66], allowing the observer to discriminate the correct color.

However, Walraven, Benzschawel, and Rogowitz [WBR87] and Arend [Are91] have demonstrated that color constancy is not really very constant. Even though most computational models provide a model for perfect color constancy, see Figure 5.7.

5.5 CONTEXT DEPENDENCE

Land and McCann's Retinex Model demonstrates the dependence of the whole visual scene on the appearance of any part. It shows the role of luminance distribution in determining the perceived hue.

In addition to physical and lower level perceptual factors influencing color perception, higher-level interpretation plays an important role. For example, the meaning of the object and the scene can affect the actual color perception; a patch of color will appear more saturated if the shape and the color are synergistic, as in a red apple or a green tree, as compared to an abstract red or green color patch.

5.6 TEMPORAL CHROMATIC EFFECTS

There is also an interaction between temporal effects and color perception. Alternating isoluminant stimuli will fuse perceptually at very low temporal frequencies, producing additive color mixture. Thus, chromatic mismatches between successive frames of a motion picture or a video are very difficult to detect, which provide the substrate for success of field-sequential color systems. Temporal luminance mismatches, however, are very easy to detect.

Conversely, one can induce the perception of color with a flickering black-and-white stimulus. The resulting perceived hue will depend on the temporal frequency ("Benham's Top"). The hypothesis is that there exists a temporal coding of color information, which is being triggered by the flickering stimulus.

Spatial-temporal interactions affect perceived color in that the sensitivity to sharp edges (high spatial frequency) decreases as a function of temporal frequency. Since sharp luminance edges enhance perceived hue, moving video images look less saturated than still images.

5.7 SUMMARY AND NOTES

We have discussed the influence color has on a number of higher level visual tasks. In particular, we examined color's help (and hindrance) to visual search, the interaction between visual and verbal activity on color perception and naming, color discrimination and color naming relationships, color contrast and constancy, context dependence, and temporal chromatic effects.

PART II

APPLICATIONS IN GRAPHICS, VISUALIZATION, AND MUTLIMEDIA APPLICATIONS

6

COLOR DEVICES

6.1 INTRODUCTION TO COLOR CALIBRATION

One of the main problems with color displays is that they vary in the way they produce colors. Not only are electronic display devices different from hardcopy devices, but even within the same category there are differences among them.

This means that without any additional intervention, color images will appear different when produced on different devices. The desire to have colors look identical independent of the device they are produced on is referred to as *device independent color*; the intervention that is expected to get us closer (though often not all that close) to this goal is referred to as *color calibration.*

6.1.1 Device independent color

Definition

By device independent color we mean creating color images that look comparable when displayed on a CRT display, projected, or printed on a printer.

Steps to device independent color

To accomplish device independent color several steps need to be taken :

1. Calibrate output devices: obtain the luminance values of the three guns for each produced color, so as to be able to reproduce them.

2. Transform to an appropriate color metric: generate a sample set of colors in a color space that represents the device using the luminance values obtained in the previous step. The choice of color space may be very important.

3. Measure the perceptual color coordinates of each of the colors in the sample set: use a spectrophotometer or tricolorimeter to obtain measurements.

4. Establish a functional relationship between the perceptual coordinates obtained in the previous step and the device driving values that generate them. Use a numerical fitting method. An appropriate ordered interpolation will usually do; 3D spline fitting may be more appropriate for such a task than linear interpolation.

We discuss some more details of these steps in turn now.

Output Device Calibration

Output device calibration is essential because it is not sufficient to characterize an image in terms of the digital-to-analog converter's (DAC) values to the red, green, and blue guns. One must measure the luminance response function for each gun. Measuring these functions is quite straightforward. Output device calibration offers the following benefits:

- It facilitates the expression of image chromaticities in standard metric spaces.
- It allows manipulating colors in a systematic fashion.
- It protects from the dangerous practice of using uncalibrated RGB values; if the R, G, and B color values are DAC values rather than luminance values, transformations from the RGB space to any metric space will be uninterpretable, and color coordinates will not be reproducible on any other device.

Selecting the Correct Color Space

Several approaches can be used to select the right color space to represent the colors in the calibration process.

One is based on colorimetric approaches—that is, on color matching data. Here, data collected by the International Commission on Illumination—known by its French name and acronym *Commission Internationale De L'Eclairage (CIE)*—such as the CIE 1931 (XYZ) data, and linear transformations of it are used, but, taking luminance into account (which the original data does not). The results are spaces for additive and subtractive color devices, see Chapters 2 and 3.

Another approach utilizes "Uniform Color Spaces." These are spaces that tend to provide better relationship between numerical changes and perceptual ones. Some of those, such as the CIELUV and CIELAB spaces, are also based on colorimetry. These two incorporate Opponent Process mechanisms. Other spaces, like the Generalized Lightness, Hue, and Saturation family of models, utilize mathematical formulation of pseudo perceptual spaces to obtain uniformity, as described in Chapters 3 and 4. Yet others, such as Guth's model, [Gut89], incorporate post-receptor processes and gain control to obtain a better perceptual relationship.

Several other approaches are based on color-naming approaches and color constancy. For more details on some of these approaches see Chapters 2, 3, and 4.

Transforming Color Information from One Device to Another

The transformation of color specifications among devices is difficult because of the *gamut mismatch* problem, which has two components:

1. Devices operate at different luminance levels, and the range of hues that are can be produced on each device depends on those luminance levels.
2. The color pigments or phosphors of each device create a range, or gamut, of colors that is not fully overlapping with the gamut of any other device.

Approaches to solving the gamut mismatch problem include pixel-by-pixel chromaticity matches, selective color preservation based on image contents, color name preservation, or color constancy preservation, i.e., appearance-preserving uniform transformations.

"Best" matches can be closest in hue, saturation, lightness, or some combination of those. That decision, as well as decisions about what metric space determines optimality, and what combination rule best describes the fit, are the responsibility of the implementor.

The answer to such questions has to be influenced by psychophysics experiments.

6.1.2 Modeling and calibration

We have described qualitatively and mathematically how color display processes work (see Chapters 1, 2, and 3).

However, correct calibration requires a more precise quantification of these processes, which we now outline.

Quantifying how color display processes work

To precisely describe a color display process, one must know

1. What colors a display device can produce (its gamut).
2. How to produce any required color ($\mathbf{c}$) in device-independent coordinates.
3. How colors produced vary over time, temperature, humidity, etc.
4. How colors vary under different lighting conditions.

Once these have been established, one can generate device driving signals $\mathbf{j}$ for any specified color $\mathbf{c}$. Those will be RGB values (additive; for light emitting displays), or CMY(K) values (subtractive; for reflecting displays, i.e., hardcopy devices).

The next step is to build a device "model," i.e., find $\mathbf{c} = f(\mathbf{j})$ and then invert f to generate $\mathbf{j}$ for any desired $\mathbf{c}$.

Device modeling approaches include physical process models, where one tries to find $\mathbf{c} = f(\mathbf{j})$ by characterizing every step $f_1, f_2, \ldots, f_n$ in the process to find f, and then invert every step f_i in the sequence to find f^{-1}.

Alternatively, one can use numerical models. This is usually the recommended approach when f is not simple, and cannot be found easily. In such case, one tries to generate sets of samples, measure the values of **c** for known values **j**, then find an approximation to f^{-1} using numerical techniques (such as, polynomial regression, splines, or hybrid interpolation).

It might be advantageous to use adaptive approaches to improve the resulting models.

The precise procedure for utilizing a numerical model requires the following steps:

1. Disable any existing lookup table (LUT) compensations and any other corrections.
2. Establish a stable gray axis of quantized levels (for consistency check, printer's control, and any other quality control).
3. Generate a sample set for the specified driver values **j** (for example, constant size chips, with certain constant R, G, B increments).
4. Measure **c** for these samples.
5. Apply the model.

To refine the model, repeat the process with a sample set equally spaced in perceptual coordinates. One can apply different weighting to colors that are considered to be more important for a given purpose (e.g., the gray axis). One can also create—and use—different models that are optimized for different tasks.

Examples of devices, model approaches and results

To model a color CRT display using a physical process, $\mathbf{c} = f_1(\ell) = f_1(f_2(\mathbf{v})) = f_1(f_2(f_3(\mathbf{j})))$, where f_1 is derived from phosphor chromaticities and the white point; f_2 is derived from the "gamma" function or is a measured relation between the lightness (ℓ) and voltage ($\mathbf{v}$) for each gun; and f_3 is obtained from the Digital to Analog Converter (DAC) values. The resulting gamut shape will depend on the CRT adjustments, and other physical conditions. For examples, see, e.g., [Rob85, Rob88].

To establish a model of a color photographic process, one has to consider complex multi-dye-layer models, which can be numerical, polynomial or spline models, but tend to be fairly linear, i.e., they do not introduce discontinuities. The results depend, among other factors, on the viewing conditions and the processing stability. For examples of resulting gamut shapes, and for more detailed discussions see, e.g., [Rob85, Rob88].

Modeling color laser printers is a complex process that is difficult to characterize (it requires modeling dyes, charge deposition, and other chemical processes). The behavior of laser printer numerical models is not very "linear." Better numerical modeling tools are needed to obtain better modeling. The results depend on viewing conditions and the laser printer environment (e.g., temperature, humidity). See [Rob85, Rob88] for more details.

6.2 GAMUT MATCHING

Once the devices in use have been modeled, and we have obtained some description of their gamuts, it is possible to attempt matching them.

Gamut matching requires the transformation of gamut-mismatches (which are colors within the gamut of a source device, which fall outside the gamut of the destination device) into colors within the gamut of the destination device. Occasionally, the only practical approach is to drop those colors, which might degrade the color quality of the image on the target device.

The mapping between devices requires understanding device limitations, color space limitations, computational limitations, and application considerations. Additionally, it is necessary to establish criteria for mappings.

Device limitations to be considered include the relationship between the device processes and their gamut shapes, such as the degree of cross-talk, the location of primary and secondary points, regions of greatest extent, and regions of quantization. Additional considerations have to deal with device sensitivities to variations, both local and global, illumination dependence, metamerism, and environmental factors.

Color space limitations include the dimensionality of the space—a full three-dimensional space as compared to two-dimensional cross sections, the gamut it covers, its perceptual properties, and known non-linearities, such as dark regions and saturation peaks and troughs.

Computational limitations include the ability to model the device characteristics, including non-linear behavior, unstable behavior, noise in measurements, and requirements for non-uniform fit (such as fitting of the gray axis); the ability to model metamerism; criteria for fit optimization; and interpolation considerations.

Special attention needs to be paid to the treatment of out-of-gamut colors. There are several approaches to handling those colors who fall outside the gamut of the destination device.

One approach treats out-of-gamut colors separately from the rest of the gamut. Under such a scheme, these colors can be assigned to prespecified values or to closest points (without or under constraints).

Alternatively, the entire destination gamut can be compressed, either linearly or non-linearly. Non-linear compression typically favors compression close to gamut edges, while leaving the rest of the gamut relatively unchanged. The nature of compression can be determined locally or globally.

Finally, adaptive mapping approaches monitor the out-of-gamut colors, and modify the gamut mapping based on the severity of their "offense."

Application considerations may further affect the mapping. These usually include constraints to maintain certain perceptual properties. The most common properties to be preserved are the relationship to primaries and secondaries, relationship to the axes, preservation of lightness, hue, or saturation (usually at maximum), and color name preservation.

6.3 SUMMARY AND NOTES

We have addressed the desire—and need—to maintain compatible colors across devices, referred to as device independent color. We have discussed the basic steps of color calibration, which is required in order to achieve device independent colors.

Finally, we have outlined the various approaches to handling the gamut mismatch problem.

For an inexpensive way to calibrate a monitor see [Cow83]. For a discussion on the mapping of gamut and printing of digital color images see [SCB88]. For general guidelines for the effective use of color in visual displays see [TM86, TS90]. For specific physiological principles see [Mur84c]. For specific perceptual principles see [Mur84b]. And for cognitive principles see [Mur84a].

7

COLOR SCALES FOR IMAGE DATA: DESIGN AND EVALUATION

In this chapter[1] we discuss the use of color to represent a parameter in a two-dimensional image. We state desirable properties of color scales. We introduce the notion of an optimal color scale, and describe the development of a particular optimal color scale; we state restrictions on the order of colors in an optimal color scale, and we present an algorithm to search for scales that obey those constraints. We briefly discuss the linearization of color scales; Chapter 8 goes into the details. We present the result of such optimization-linearization process in the form of the *Linearized Optimal Color Scale (LOCS)*. We describe observer performance experiments to evaluate the merits of color scales for image data. The evaluations show that observers perform somewhat better with the developed LOCS than with a previously advocated scale, the heated-object color scale, but they perform significantly better with a linearized gray scale than with either of the color scales. We discuss possible reasons for this result.

7.1 INTRODUCTION

Color scales have been used for the representation of single-parameter distributions for quite some time. Indeed, every time an image of some parameter distribution is displayed, some color scale has been employed. Most often, it is the gray scale (which is not considered a color scale by most of us). The main advantages of the gray scale are its simplicity and the natural sense of order that it induces. Its main disadvantage are a limited *perceived dynamic range* (only in the order of 60 – 90 just-noticeable-differences—JNDs). This may cause a problem whenever the range of data values represented is larger than these numbers (which is very often the case).

[1] This chapter is adapted from [LH92].

We discuss the representation of a parameter using a color scale. In Section 7.2, we state desirable properties of color scales. Section 7.3 presents a brief description of commonly used scales in computer graphics applications. In order to attain the desirable properties, we introduce in Section 7.4 the notion of an *optimal color scale*, and outline the general problem of finding an optimal scale. Section 7.5 presents a specific solution approach and implementation, including the algorithm *OPTIMAL-SCALES*. The results of *OPTIMAL-SCALES*, which yield a particular optimal color scale, are presented in Section 7.6. The linearization of color scales is briefly discussed in Section 7.7, where the results of the two-step process are presented in the form of the Linearized Optimal Color Scale (LOCS); detailed discussion of the linearization process is postponed till the following chapter, 8.. Observer performance experiments to evaluate the merits of color scales for image data are described in Section 7.8. The results of these experiments are presented in Section 7.9. They show that although observers performed somewhat better with the newly developed LOCS than with the previously advocated heated-object color scale, they performed significantly better with a linearized gray scale than with either of the color scales. In Section 7.10 we discuss possible reasons for these results and some alternate solutions that may perform better. We also present some additional observations about the potential contribution of color scales for the perception of information in images.

7.2 DESIRED PROPERTIES FOR COLOR SCALES

Given a sequence of numerical values $\{v_1 \leq \ldots \leq v_N\}$, which are to be represented by the colors $\{\mathbf{c}_1, \ldots, \mathbf{c}_N\}$ respectively, we identify three desired properties.

7.2.1 Order

The colors used to represent the values in the scale should be perceived as preserving the order of the values. The relationship among the colors should be

$$\mathbf{c}_1 \textit{ perceived-as-preceding } \ldots \textit{ perceived-as-preceding } \mathbf{c}_N .$$

For example, representing a temperature scale, one can use the notions of "cold" and "warm" colors and their proportional mixtures to convey a scale from cold to hot temperatures.

7.2.2 Uniformity and representative-distance

The colors used should convey the distances between the values that they are representing. Colors representing values that are equally different from each other along the scale should be perceived as equally different from each other. That is, for any $1 \leq i, j, m, n \leq N$, if $v_i - v_j = v_m - v_n$ then we would like to have $pd(\mathbf{c}_i, \mathbf{c}_j) = pd(\mathbf{c}_m, \mathbf{c}_n)$, where $pd(\mathbf{c}, \mathbf{c}')$ is the *perceived distance* between the colors $\mathbf{c}$ and $\mathbf{c}'$. Important differences in the values should be represented by colors clearly perceived as different, while close values should be represented by colors that are perceived to be close to each other. For example, representing flow information, one can use complementary colors to represent flows in opposite directions and similar colors to represent flows in the same direction but at slightly different angles.[2]

7.2.3 Boundaries

The color scale should not create perceived boundaries that do not exist in the numerical data. That is, it should be able to continuously represent continuous scales.

7.3 COMMONLY USED SCALES

The most commonly used scale is the *gray scale.* While not considered a color scheme, it is the result of traversing the color solid along the achromatic (lightness) axis. This can be implemented by keeping equal intensities for the three primaries red, green, and blue (R, G, B) and increasing them monotonically from 0 to M—the maximum they can assume. See Figure 7.1 and 7.2 for a non-linearized and a linearized version, respectively.

7.3.1 The Rainbow scale

The *rainbow scale* (Figure 7.3) is so called because it traverses the color solid along a path from black to white, passing through all the hues of the rainbow (Red, Orange, Yellow, Green, Blue, Indigo, Violet—ROY-G-BIV), though at different lightnesses. Other variations of this very popular (but not so very effective) scale maintain constant lightness throughout the scale.

[2]This property represents the unification of the notions of *associability* and *separation* presented in [PZJ82] and [RO86], respectively.

7.3.2 The Heated-Object and the Magenta scales

These two scales traverse the color solid from black to white following two different paths starting at the red axis. Both scales are based on the claim (which has never been confirmed) that natural color scales seem to be produced when the intensity of the three primary colors red, green, and blue rise monotonically and with the same order of magnitude of intensities throughout the entire scale [PZJ82]. Both scales satisfy this property.

The *heated-object scale* is implemented by increasing the gun intensities in the order red, green, blue, see Figure 7.4. Its path through the color solid is limited to 60° clockwise from the red axis. It is based on the fact that the human visual system has maximum sensitivity to luminance changes for the orange-yellow hue. For example, in the Munsell Color System there are more distinct divisions for high-value highly-saturated yellows than there are for other hues [Mun05, Mun69, Mun76].

The *magenta scale* is implemented by increasing the gun intensities in the order red, blue, green, see Figure 7.5. Its path through the color solid is limited to 60° counter-clockwise from the red axis. It is based on the fact that the human visual system is most sensitive to hue changes for the magenta hue. For example, it is reported in [Mac43] that experiments performed to measure discrimination of hue showed that the best hue discrimination is achieved for the purples.

7.4 THE PROBLEM OF OPTIMAL COLOR SCALES

A *color scale* is a pictorial representation of numerical values in which each value is assigned its own color. Since many of the colors in a scale may not be *perceived* as distinct, for an informative representation it is important to maximize the number of distinct perceived colors along the scale (referred to as the number of *just noticeable differences—JNDs*).

We are mainly concerned with the representation of numerical information on color monitors and so we represent each color by a three-dimensional vector (r, g, b) of the intensities of the red, green, and blue guns needed to generate that color on the screen. Denoting by M the maximum intensity for any of

Figure 7.1 A non-linearized grayscale. Black increasing to white, and their gun intensities. The abscissa represents the progression of the colors along the scale. (See Web site for color version.)

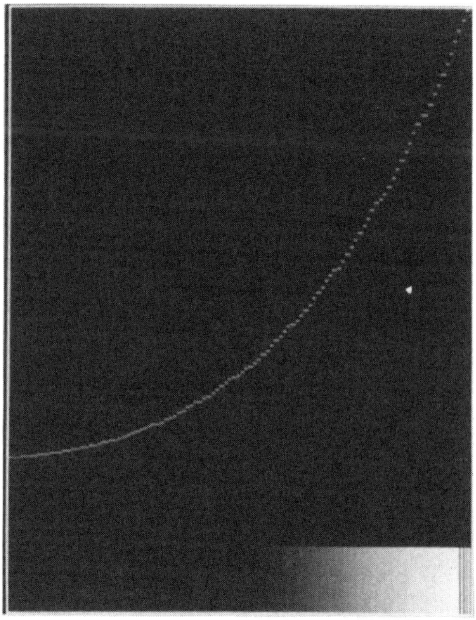

Figure 7.2 A linearized grayscale. (See Web site for color version.)

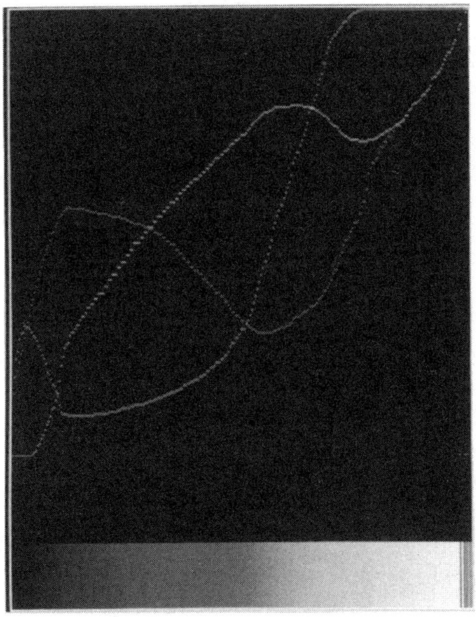

Figure 7.3 The rainbow scale. (See Web site for color version.)

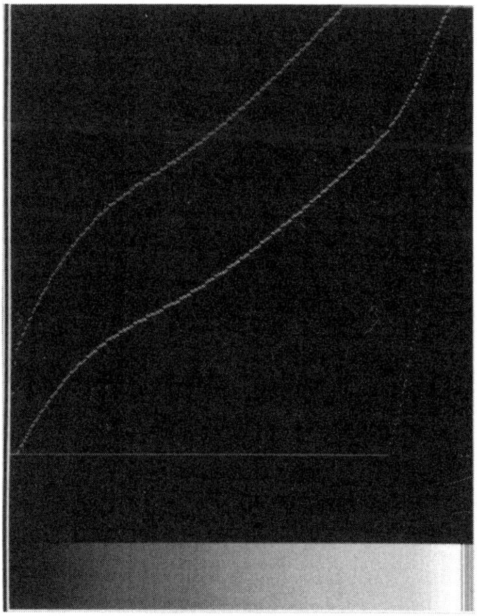

Figure 7.4 The heated-object scale. (See Web site for color version.)

Figure 7.5 The magenta scale. (See Web site for color version.)

the primaries, the set of vectors (r, g, b) such that $0 \leq r, g, b \leq M$ is called the *colorcube* or *RGB space*. The colorcube has been discussed in detail in Chapter 3.

Because the human observer is sensitive to changes in lightness, a sense of order is achieved by color scales that start with a dark color for the smallest value, and end with a light color for the largest value, going through colors with increasing lightness. The changes in the hue and the saturation of the colors in the scale also influence the naturalness of the order of colors in the scale.

We seek an *optimal scale*—one that maximizes the number of JNDs while maintaining a natural order among its colors. Naturalness can be enforced by restricting the scale in various ways [Lev88, LH87b]. For example, as already mentioned, it has been claimed that scales that preserve the order of magnitude of the primaries r, g, and b appear to induce a natural order [PZJ82].

In the next section we state the precise restrictions we impose on a scale meet our criteria for it to be an optimal color scale. We use the same definitions of lightness, hue, and saturation we presented in Chapter 3. Our general approach, however, is applicable to other specific definitions of the intuitive concepts [FvD82, Lev88, LH87a, LH87b, Smi78a].

7.4.1 Restrictions

1. We assume that the numerical scale we wish to represent is discretized into a fixed number, N, of equidistant increasing values, denoted by $v_1, \ldots, v_N$. Correspondingly, it is only necessary to construct a color scale that consists of N colors, $\mathbf{c}_1, \ldots, \mathbf{c}_N$, uniformly spaced perceptually (but not necessarily equidistant in color space), see Section 7.2.2. (While, initially, we formulate the problem such that the colors $\mathbf{c}_i$, $i = 1, \ldots, n$ are distinct, this restriction may have to be relaxed later. This will depend on the relationship between the number of colors, N, in the scale and the number of intensity levels, M, available on the monitor.) We denote $\mathbf{c}_n$ by the vector of primaries (r_n, g_n, b_n), for $1 \leq n \leq N$.
2. In order to maximize the number of JNDs it is reasonable to demand that $\mathbf{c}_1$ be black—$(0, 0, 0)$ and $\mathbf{c}_N$ be white—(M, M, M).
3. Since, as mentioned in Section 7.2.1, we want to preserve the naturalness of ordering, we insist that, for $1 \leq n \leq N-1$,

$$r_n \leq r_{n+1}, \qquad g_n \leq g_{n+1}, \qquad b_n \leq b_{n+1},$$

and

$$r_n + g_n + b_n < r_{n+1} + g_{n+1} + b_{n+1}.$$

This restriction might need to be relaxed if due to the number of intensity levels available on the monitor the final colors in the scale are some approximation of the originally selected colors.

4. We do not wish to mix achromatic and chromatic colors in the scale as this might affect our ability to meet the desired properties of Section 7.2, in particular uniformity (see Section 7.2.2) and boundaries (see Section 7.2.3). Thus, we insist that if $\mathbf{c}_n$ is achromatic (i.e., $r_n = g_n = b_n$), for any n, $1 < n < N$, then $\mathbf{c}_n$ is achromatic for all n and the scale is a *grayscale*. Otherwise, $\mathbf{c}_n$ has a well-defined hue h_n for $1 < n < N$. In this case, to preserve the naturalness of order, we insist that for some m, $1 < m < N$, either

$$h_2 \leq \ldots \leq h_m < h_{m+1} + 360 \leq \ldots \leq h_{N-1} + 360,$$

or

$$h_2 \geq \ldots \geq h_m > h_{m+1} - 360 \geq \ldots \geq h_{N-1} - 360.$$

Furthermore, we also investigate optimal color scales under different restrictions on the total range of hues as this may affect the desirable properties of Section 7.2. In any case, we do not allow the sequence of hues to wrap around the achromatic axis (i.e., the total angle covered by moving from h_2 to h_{N-1} must be less than 360°).

5. To maintain natural order and to avoid artificial boundaries, we also desire the saturations $s_2, \ldots, s_{N-1}$ of the colors $\mathbf{c}_2, \ldots, \mathbf{c}_{N-1}$ to be monotonic. In the case of the chromatic scales, these values are strictly greater than zero by the condition described in 4. On the other hand, by the condition described in 2, $s_1 = s_N = 0$. These requirements impose that either

$$s_1 \leq s_2 \leq s_3 \leq \ldots \leq s_{N-1},$$

or

$$s_2 \geq s_3 \geq \ldots \geq s_{N-1} \geq s_N.$$

7.4.2 The maximization problem

Under the given restrictions, we want to maximize $j(\mathbf{c}_1, \mathbf{c}_2) + \ldots + j(\mathbf{c}_{N-1}, \mathbf{c}_N)$, where $j(\mathbf{c}, \mathbf{c}')$ is a measure of the change in terms of JNDs resulting by moving from the color $\mathbf{c}$ to the color $\mathbf{c}'$. Our JND information is based on the CIELUV *Uniform Color Space (UCS)* [CIE78]. Each color $\mathbf{c} = (r, g, b)$ in RGB space has a unique representation $(L^*(\mathbf{c}), u^*(\mathbf{c}), v^*(\mathbf{c}))$ in CIELUV. Details of the transformations can be found in Chapter 2 and [CIE78, JW75, Lev88, LH87b, Taj83a, Taj83b]. Thus, our problem becomes the following. Find $\mathbf{c}_1, \mathbf{c}_2, \cdots, \mathbf{c}_N$ satisfying certain restrictions such that

$$\sum_{n=1}^{N-1} \sqrt{(L^*(\mathbf{c}_{n+1}) - L^*(\mathbf{c}_n))^2 + (u^*(\mathbf{c}_{n+1}) - u^*(\mathbf{c}_n))^2 + (v^*(\mathbf{c}_{n+1}) - v^*(\mathbf{c}_n))^2} \tag{7.1}$$

is maximum. Some of the restrictions we use are given above, other (more specific) ones are stated below.

7.5 SOLUTION APPROACH AND IMPLEMENTATION

We first state some more specific restrictions and definitions. These are mainly determined by compatibility with commonly available equipment and by the scales with which our scale is to be compared.

We only allow colors $\mathbf{c} = (r, g, b)$ in which r, g, and b are integers between 0 and M (on a typical monitor $M = 255$). This is merely a matter of convenience since it enables us to implement and use the scale easily using existing equipment. Since we want to compare our "optimal" color scales with the gray (achromatic) scale, for the initial phase of the implementation, we set $N = M + 1 = 256$. At the later phase of linearization, the actual number of distinct colors along the scale may decrease as approximations are computed to achieve equally perceivable steps that are realizable with the selected value of M. This should be sufficient since, based on psychophysical experiments with other scales, one expects the number of JNDs along the scale to be less than 256 [PZJ82].

The lightness, hue, and saturation of a color specified within the colocube are determined using the LHS-triangle model, which is defined by Equations 3.1, 3.8, 3.11, 3.12, 3.13 of Chapter 3. The specific selection of the LHS model is a choice of convenience; we have shown in Chapter 3 that all the LHS models currently used in computer graphics are special cases of the general GLHS family. We find the coordinate transformations from the colorcube to the LHS-triangle model to be conceptually the most straightforward.

We organize all the points in the $(M+1)^3$ discretized colorcube in the order of their lightnesses, which we define by the lightness function of the LHS-triangle

j	0	1	2	3	4	5	6	7	8	9
Δ_r	3	2	2	1	1	1	0	0	0	0
Δ_g	0	1	0	2	1	0	3	2	1	0
Δ_b	0	0	1	0	1	2	0	1	2	3

Table 7.1 The $\Delta(j)$ operators.

model given in Equation 3.1 in Chapter 3. For each n, $0 \leq n \leq N-1$, this yields a set of colors

$$C(n) = \{(r,g,b) : r,g,b \text{ are integers between 0 and M and } (r+g+b)/3 = n\}. \quad (7.2)$$

For the initial phase of implementation we insist that $\mathbf{c}_n \in C(n-1)$, for $1 \leq n \leq N$, which is a further restriction to what was stated before . This restriction is not a property that we will require the final scale to have, but it simplifies the construction of that scale.

The requirement that $\mathbf{c}_n \in C(n-1)$ for all n implies that $\mathbf{c}_2 \in C(1)$, i.e., $\ell(\mathbf{c}_2) = 1$. The only colors that can have lightness 1 are the achromatic color $(1,1,1)$ and colors for which $min(\mathbf{c}) = 0$, which means, by definition of the saturation function that their saturation $s(\mathbf{c}) = 1$. Thus, the monotonicity restriction on the saturation can only be fulfilled if the saturation along the scale is monotonically non-increasing. By the symmetry of the colorcube with respect to $\mathbf{c}_1$ and $\mathbf{c}_N$, the monotonicity restriction on the saturation can only be fulfilled if the saturation along the scale is monotonically non-decreasing. Thus, the monotonicity requirement implies that, for a chromatic scale, $s_2 = \cdots = s_{N-1} = 1$.

To construct a scale in which $\mathbf{c}_n \in C(n-1)$ for $1 \leq n \leq N$ and each gun is independently non-decreasing, we define a set of operators

$$\Delta(j) = \{(\Delta_r(j), \Delta_g(j), \Delta_b(j)) : j = 0, \ldots, 9\}, \quad (7.3)$$

such that the lightness increase for $j = 0, \ldots, 9$ is

$$\Delta L = \frac{\Delta_r(j) + \Delta_g(j) + \Delta_b(j)}{3} = 1. \quad (7.4)$$

The operators are given in Table 7.1. From any given color $\mathbf{c}_n = (r_n, g_n, b_n)$ we generate the set of all possible extensions to the scale by applying each of the operators $\Delta(j)$, $j = 0, \ldots, 9$ to $\mathbf{c}_n$. A permissible color scale can progress to a newly generated point only if the new point fulfills all the restrictions on the behavior of the scale given above.

From all the permissible paths that can be generated in this way, we are interested in finding those that maximize the number of JNDs. To find them, we used *dynamic programming* [AHU76]. In order for dynamic programming to be feasible, it is essential that the permissibility of a scale at any given point be a local decision.

We translate our problem into one of finding all the optimal paths in a set of directed graphs defined as follows. For each set of starting hue h_s, ending hue h_e, and hue-progression direction dir, we define a directed graph $G = (V, E)$ such that the set of nodes is

$$V = \{(r, g, b) : 0 \leq r, g, b \leq M,\ r, g, b \text{ integers}\}, \tag{7.5}$$

which is the set of $(M + 1)^3$ colors in the discretized colorcube, and the set of directed edges is

$$E = \{(\mathbf{c}, \mathbf{c}') : \mathbf{c}' = \mathbf{c} + \Delta(j), j \in 0, \ldots, 9, \text{ and } PERMISSIBLE(\mathbf{c}, \mathbf{c}')\}, \tag{7.6}$$

where $PERMISSIBLE(\mathbf{c}, \mathbf{c}')$ means that the step from $\mathbf{c}$ to $\mathbf{c}'$ can occur in some permissible color scale. Having defined the graphs as such, one can prove that any directed path from black ($\mathbf{c}_1 = (0, 0, 0)$) to white ($\mathbf{c}_N = (M, M, M)$, recall $N = M + 1$) through such a directed graph is a permissible scale. Note also that once h_s, h_e, and dir are fixed, the truth of $PERMISSIBLE(\mathbf{c}, \mathbf{c}')$ can be locally determined for any $\mathbf{c}$ and $\mathbf{c}'$. For any color $\mathbf{c}$ in the cube, Algorithm *OPTIMAL-SCALES* (see Figure 7.6) produces the set of optimal paths $P(\mathbf{c})$ and the optimal color distance $d(\mathbf{c})$ between *black* and $\mathbf{c}$. In particular, since *white* ($\mathbf{c}_N$) is a color, *OPTIMAL-SCALES* produces $P = P(\mathbf{c}_N)$ and $d(\mathbf{c}_N)$.

7.6 RESULTS OF *OPTIMAL-SCALES*

Given the six hue sectors defined by Eq. 3.12, there are 72 different choices[3] of the starting hue h_s, the ending hue h_e, and the hue-progression direction dir:

[3] This is under the restriction that scales start and end at the beginning of sectors. If this restriction is removed, there are 162 different choices: 9 (start) × 9 (end) × 2 (up/down), see Table 7.1. However, the behavior of the scales is adequately characterized under this restriction, which simplifies the search.

Algorithm *OPTIMAL-SCALES*
Input: M—which defines the discretized colorcube V and the length of the scale $N = M + 1$, h_s—the starting hue, h_e—the ending hue, and *dir*—the hue-progression direction.
Output: $d(\mathbf{c}_N)$—the optimal color distance from black to white and P—the set of optimal scales from black to white.
Auxiliary Variables: For every $\mathbf{c} \in V$, $d(\mathbf{c})$ is a number and $e(\mathbf{c})$ is a set of elements in V.

begin
  for each $\mathbf{c} \in V$ **do begin**
    $d(\mathbf{c}) := 0;$
    $e(\mathbf{c}) := \emptyset$
  endfor;
  for $n := 1$ to M **do**
    for each color $\mathbf{c} \in C(n-1)$ **do**
      for $j := 0$ to 9 **do begin**
        $\mathbf{c}' := \mathbf{c} + \Delta(j);$
        if *PERMISSIBLE*$(\mathbf{c}, \mathbf{c}')$ **then begin**
          if $d(\mathbf{c}) + j(\mathbf{c}, \mathbf{c}') > d(\mathbf{c}')$ **then begin**
            $d(\mathbf{c}') := d(\mathbf{c}) + j(\mathbf{c}, \mathbf{c}');$
            $e(\mathbf{c}') := \{\mathbf{c}\}$
          endif;
          if $d(\mathbf{c}) + j(\mathbf{c}, \mathbf{c}') = d(\mathbf{c}')$ **then** $e(\mathbf{c}') := e(\mathbf{c}') \cup \{\mathbf{c}\}$
        endif
      endfor
    endfor
  endfor;
  $P := \{(\mathbf{c}_1, \mathbf{c}_2, \ldots, \mathbf{c}_N) : \mathbf{c}_N = (M, M, M)$ and for $1 \leq n \leq M,\ \mathbf{c}_n \in e(\mathbf{c}_{n+1})\};$
end; {*OPTIMAL-SCALES*}

Figure 7.6 Algorithm *OPTIMAL-SCALES*.

$h_s \in \{0, 60, \ldots, 300\}$; $h_e \in \{0, 60, \ldots, 300\}$, such that the number of sectors the scale covers is in $\{1, 2, \ldots, 6\}$; and $dir \in \{down, up\}$, where *down* means clockwise progression and *up* means counter-clockwise progression.

Additionally, one can replace the monotonicity requirement of the saturation with a less restrictive requirement, namely, that the saturation of any color in the scale be larger than a predefined lower limit (the least restrictive case would be $s(\mathbf{c}_n) > 0$ for $2 \leq n \leq N-1$, which only requires that all the colors except the first and the last be chromatic).

As reported by Levkowitz and Herman in [LH92], we have ran *OPTIMAL-SCALES* on a 32^3 RGB colorcube (that is, $M = 31, N = 32$)[4] with all the 72 different choices of h_s, h_e, and dir, and with several values for the lower limit of the saturation. Our results show that:

- Scales whose saturation was allowed to fluctuate non-monotonically did not induce the required natural order; colors with lower saturation were perceived to be of lesser lightness than fully saturated ones. This result means that the only acceptable color scales are those with constant or monotonic (probably non-decreasing) saturation throughout the scale. In our case we selected $s(\mathbf{c}_n) = 1$ for $2 \leq n \leq N-1$.
- Of the 72 different choices, the optimal scale with the longest distance in the CIELUV space is the scale that results from positive hue progression from 0 to 180°. After expanding it to a 256-color scale by scaling of the gun intensities and linear interpolation of missing colors, its total CIELUV-distance was 5.82 times that of the grayscale. Figure 7.7, shows the resulting optimal color scale (OCS), the gun-intensities that implement it, and a plot of the CIELUV distances between adjacent colors in the scale.
- Inspecting the scale in Figure 7.7, we notice that the overall progression of the scale is not perceived to be uniform. Indeed, the non-uniformity of the CIELUV distances between adjacent colors in the scale can be noticed in the plot in Figure 7.7, (b).

The non-uniform behavior of the CIELUV distance values explains the perceived non-uniformity in the scale. In order to correct the scale we applied a

[4] The time and space required to run *OPTIMAL-SCALES* are proportional to the size of the colorcube. We found 32 to be a reasonable size for the initial runs in that it is large enough to yield reliable results but has reasonable time and space requirements. Selected runs with the particular choices that yielded optimal scales on larger colorcubes ($N = 64, 128$) yielded results that are consistent with the smaller colorcube ones.

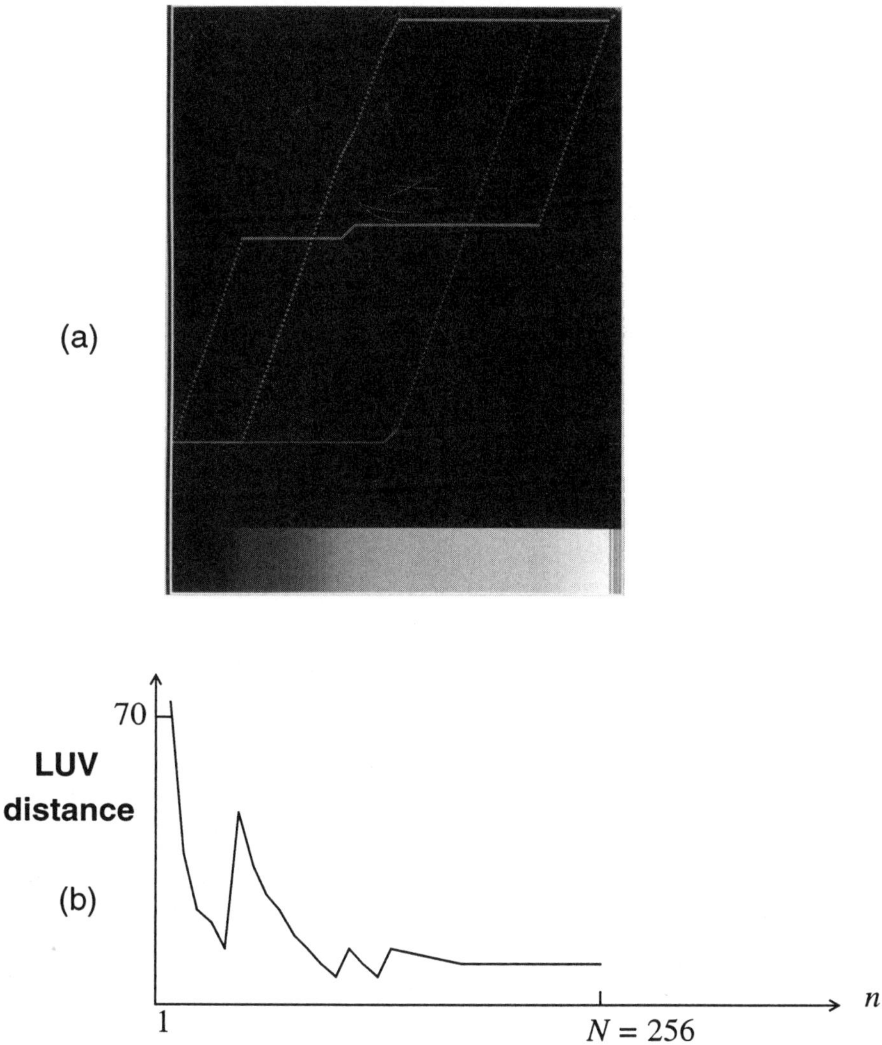

Figure 7.7 The optimal color scale resulting from $N = 32$, $h_s = 0$, $180 \leq h_e \leq 360$, and $dir = up$ and interpolation to 256 colors: (a) The RGB gun-intensities and the scale, and (b) the CIELUV distances between adjacent colors in the scale.

linearization procedure, which is described briefly in the next section and in greater detail in Chapter 8.

7.7 LINEARIZATION OF COLOR SCALES

Linearization of a color scale is a process in which additional colors are inserted in the scale in such a way that the perceived distances between adjacent colors of the extended scale are as uniform as possible. Pizer [Piz81b] and Robertson [Rob85] describe algorithms to linearize a scale. We have developed our own algorithm, which we describe in detail in Chapter 8. The main idea is to produce from the original scale a super-sampled scale (via interpolation) and to scan the super-sampled scale for a subset of colors that are equally (or almost equally) spaced perceptually.

We ran our linearization algorithm on an input optimal color scale with $N = 32$ colors. The result is a 256-color scale, which we refer to as the Linearized Optimal Color Scale (LOCS). Figure 7.8 shows the LOCS, its gun intensities, and the CIELUV distances between adjacent colors along the scale. Variations of those distances are small and, indeed, the colors in the LOCS are perceived to be uniformly distanced [Lev88].

The total CIELUV distance from black to white along the LOCS is 6.01 times that of the linearized grayscale; the total distance along the linearized heated-object scale of Figure 7.4 is 4.09 times that of the linearized grayscale. (All these distances are for scales with 256 distinct colors.)

7.8 EVALUATION OF OPTIMAL COLOR SCALES

The ultimate test for the efficacy of a color scale (optimal or else) can only be accomplish through rigorous, objective evaluation, using statistically significant sample spaces and numbers of observers. We now describe such an evaluation process.

We used the method of relative operating characteristic (ROC) [SP82] to evaluate the performance of the different scales. The specific performance criterion was the ability to detect the existence of artificially superimposed lesions in

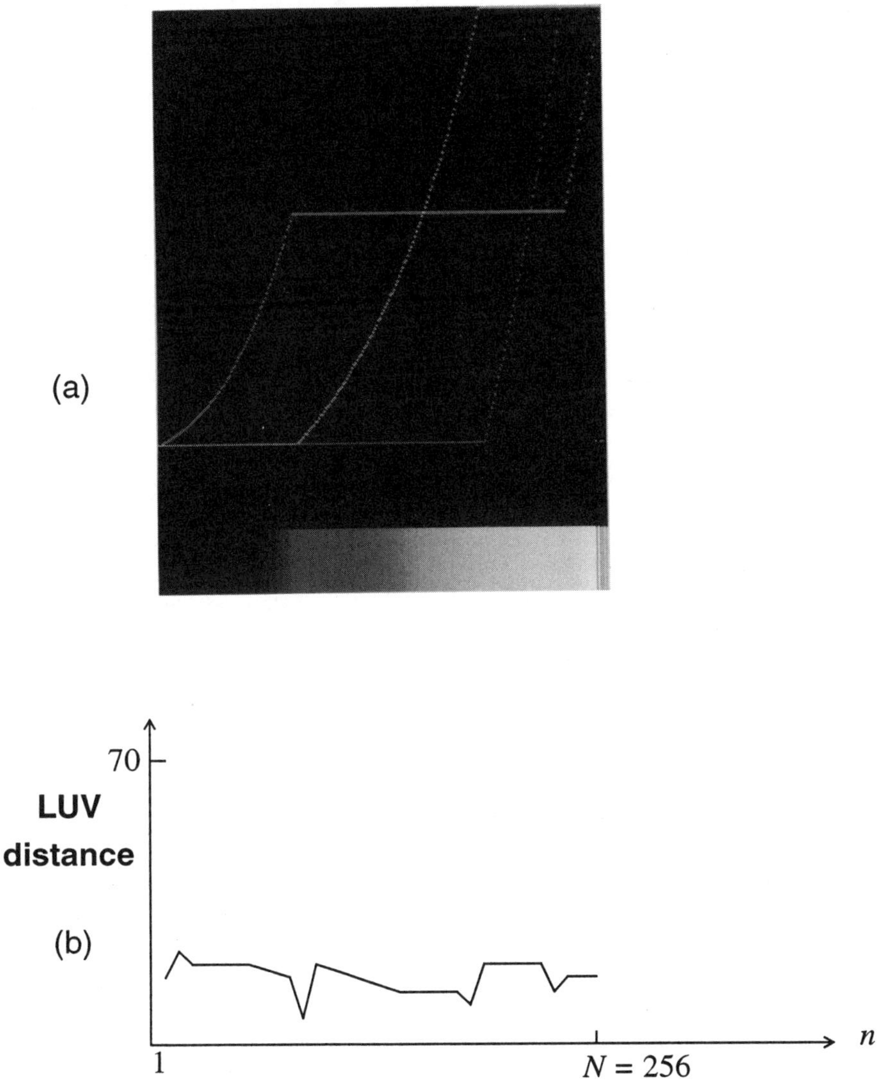

Figure 7.8 LOCS: The linearized optimal color scale resulting from $N = 32$, $h_s = 0$, $180 \leq h_e \leq 360$, and $dir = up$ and interpolation to 256 colors: (a) The RGB gun-intensities and the scale, and (b) the CIELUV distances between adjacent colors in the scale. See Chapter 8 for details.

brain cross sections. The goal of the study was to compare the performance of observers reading the same images when presented using the three scales: the linearized gray scale, the heated-object scale (linearized), and the LOCS. Details of the following brief description can be found in [Lev88].

We used in the study digitized photographs of normal brain slices. The quality of these pictures enabled us to know that the images were indeed normal. The area of the brain in the photographs was mapped into display intensities of 25–200. Uncorrelated random noise of mean zero and standard deviation eight was superimposed on each image. From the original set of 23 photographs we generated 138 different images by superimposed random noise. For the actual study, only 120 of the images were used, and 20 of these formed the learning set. A randomly-selected half of the entire data-set was used to generate the abnormal subset. Each of the images selected was superimposed (on a pixel-by-pixel basis) with a round, "bell-shaped" simulated lesion generated by additional noise of mean 16 and standard deviation 8. Figure 7.9 shows an example of an abnormal image from the resulting data-set.

The first three volunteers performed the study with this data-set. After examining the results of the first session, it appeared to us that the given data-set was somewhat easy. We generated a second data-set, with the mean for the lesions reduced from 16 to 14. Two additional readers performed the study with the second data-set.

Each reader was presented with images in the three scales in various orders. Before each new scale the reader was presented with a training session, after which he was taken through a feedback session. The first scale presented to each reader was repeated at the end for a second reading, for intraobserver comparison. The scale used to display the images was displayed alongside the images at all times. All viewing parameters were set, and the readers were not able to change them.

The observers were given an automated questionnaire on the screen. Their responses to the given questions were direct to the computer. The first response was a classification of {Normal/Abnormal}. The second was the level of confidence {Very Likely/Probably/Possibly}. The observers were of diverse backgrounds;[5] four had significant experience with computer displays of medical images (mainly black and white), and one had some exposure to such images but little experience using them. The ROC curves for one of the observers are

[5] An electrical engineer, a systems manager, a bio-engineering graduate student, a researcher in Positron Emission Tomography, and an obstetrician/gynecologist, all with a masters or a doctoral degree.

<table>
<tr><th>#</th><th colspan="2">LinGray</th><th colspan="2">Heated-Object</th><th colspan="2">LOCS</th></tr>
<tr><td>1</td><td>.921, .031</td><td>.925, .031</td><td colspan="2">.791, .048</td><td colspan="2">.799, .045</td></tr>
<tr><td>2</td><td colspan="2">.945, .027</td><td colspan="2">.882, .047</td><td>.913, .032</td><td>.910, .036</td></tr>
<tr><td>3</td><td colspan="2">.920, .029</td><td colspan="2">.836, .041</td><td>.893, .039</td><td>.811, .053</td></tr>
<tr><td>4</td><td colspan="2">.639, .058</td><td colspan="2">.478, .063</td><td>.605, .060</td><td>.652, .061</td></tr>
<tr><td>5</td><td colspan="2">.918, .030</td><td>.664, .057</td><td>NA</td><td colspan="2">NA</td></tr>
</table>

Table 7.2 The areas under the ROC curves and their associated estimates of the standard deviations as computed by the Dorfman and Alf [DA69] program for readers 1–5 for the different scales. NA means that the program failed on the data-set. Double entries indicate that that scale was the first and last for that reader.

shown in Figure 7.10. Note that a larger area under the ROC curve indicates better performance [SP82].

Two comments on the limited scope of the experiments and their results are appropriate: First, though very common in studies of medical imaging systems, pure detection techniques are limited in their scope. Most techniques, in particular in medical imaging, require in addition to detection also classification and localization. Secondly, the number of experiments run (four sessions for each of five readers) was relatively small (though ROC studies with three readers and three sessions per reader are quite common in medical imaging research).

7.9 RESULTS OF THE EVALUATION

Table 7.2 gives the values of the areas under the ROC curves of the different readers for the different scales as computed by the program of Dorfman and Alf [DA69], which also provides for each entry an estimated standard deviation. The average performance of the three readers in the higher contrast study as measured by the area under the ROC curve, is 0.929 for the linearized gray scale, 0.836 for the heated-object scale, and 0.854 for LOCS. Further analysis [Lev88] leads to the conclusion that for the task at hand LOCS appears to be better than the heated-object scale, but they are both significantly worse than the linearized gray scale.

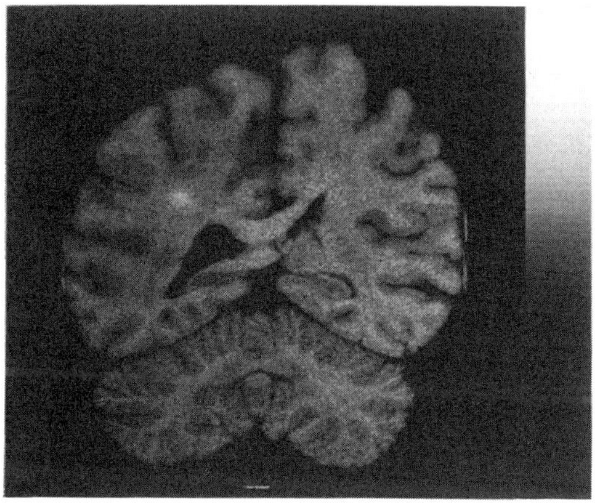

(a)

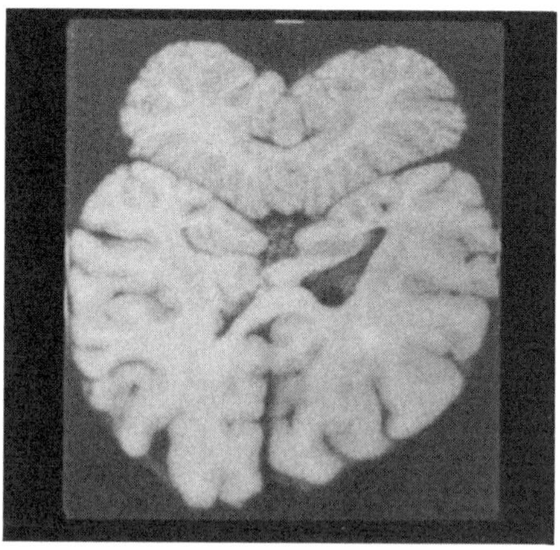

(b)

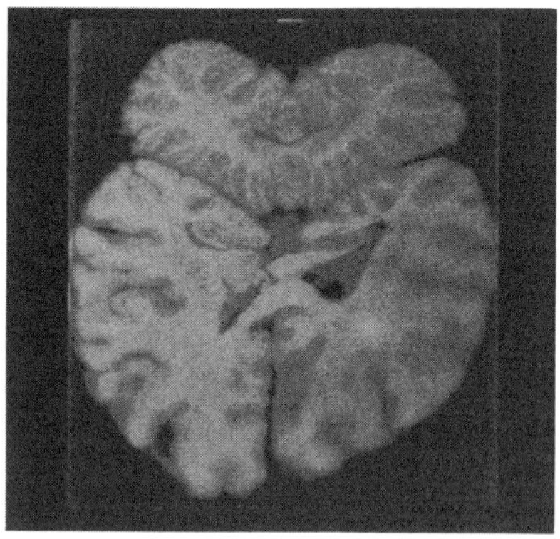

(c)

Figure 7.9 An example of an abnormal image used in the study. (a) The linearized grayscale , (b) the heated-object scale, and (c) LOCS. (See Web site for color version.)

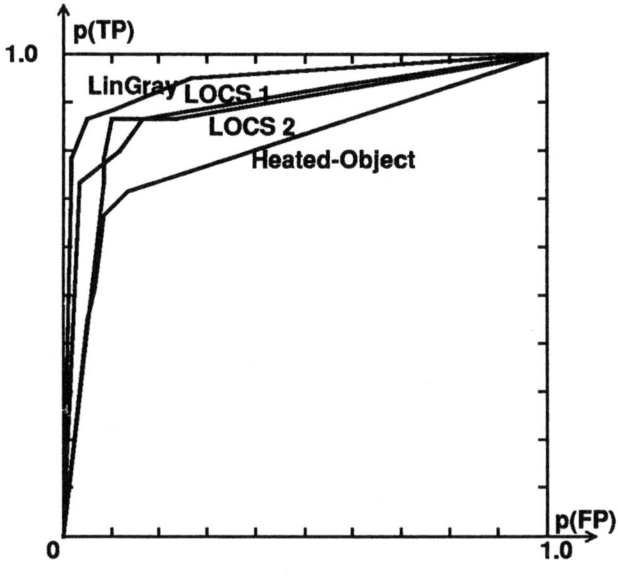

Figure 7.10 ROC curves of one of the observers.

7.10 SUMMARY AND NOTES

The superiority of the gray scale over the heated-object scale for detection of small abnormalities in a reasonably complicated background has been previously claimed in [Tod83]. The hope that by developing an "optimal" color scale one might outperform the gray scale turned out not to be so, at least for the particular task described. One reason for this may be that the perceived change in color due to a color's surround was not taken into consideration in the design of our optimization task. Other color scales, such as those proposed in [Rob85, War88] may perform better from this point of view. Another reason may be the inadequacy of the CIELUV [CIE78] space for modeling perceived uniformity. Alternate, possibly more uniform, spaces [LR86, Mun76] could be used with our optimization methodology to provide scales that may perform better. The approximating nature of the lightness function used may have some influence on the resulting scale, though no known mathematical modeling of lightness is an accurate model for human perception of lightness.

While the results of the psychophysical experiments do not show that the color scales in the study yield better performance than the gray scale, we feel that the results are limited to the particular conditions of the experiment, namely the task of blob detection. The relatively small number of runs of the test also limits the scope of the results. Through our interactions with the various scales under different circumstances we found that for each scale there were conditions under which that scale appeared to be superior to the other ones. Colleagues who participated in the studies (and others who participated in various interactions with the scales) reported similar impressions. This leads us to conjecture that the availability of a number of scales in an easily changeable way can substantially increase the perception of information in images over the use of any single scale.

To attempt alleviating some of the problems mentioned above, we are investigating extensions and generalizations to the approach presented here. The generalization approach consists of building a modular optimization system, built of five modules. Alternative choices can be "plugged in" to a module, based on the specific situation and criteria.

The modules are:

1. *Optimization objective:* "What to optimize," such as the perceived dynamic range, defined.by the number of JNDs along the scale.
2. *Optimization rules and constraints:* Such as ranges and allowed changes in lightness, hue, and saturation.

3. *Reference and measurement criteria:* Such as the reference color model (e.g., CIELUV, Munsell), and JNDs metric (e.g., Euclidean distance in CIELUV).
4. *Optimization approach:* Such as Dynamic Programming.
5. *Post processing:* Such as linearization.

Such a modular approach will allow much broader range of optimizations, which would help constructing better scales.

8

PERCEPTUAL STEPS ALONG COLOR SCALES

Color scales are commonly used to represent numerical information visually. Most scales are derived from some physical or mathematical behavior; even worse, sometimes, they are selected based solely on hardware capabilities. In most known cases, no consideration is made to the perceptual capabilities of the human observer, who is ultimately the "consumer" of the information to be delivered by the scale.

This chapter[1] presents a method and an algorithm for the derivation of color scales, such that their perceptual properties—in particular, the perceptual steps between colors along the scale—can be controlled by the scale designer. This approach has been used for the design of the linearized gray scale and the Linearized Optimized Color scale (LOCS) described in Chapter 7. These scales are demonstrated and are also compared to the linearized Heated-Object scale. [Lev96].

8.1 INTRODUCTION

A *color scale* is a pictorial representation of a set of distinct categorical or numerical values in which each value is assigned its own color. Color scales have been used for the representation of parameter distributions for quite some time.

People have been seeking alternative color scales—typically referred to as *pseudo-color* scales—for aesthetic, as well as functional or perceptual reasons.

[1] This chapter is adapted from [Lev96].

Schuchard cites several studies implying that medical images displayed with pseudo-color scales may give more information to the observer, but admits there is no conclusive evidence that proves this. He hints, with no evidence though, that the Heated-Object scale possesses some desirable qualities for use in medical image display [Sch87, Sch90b].

There is no controversy that appropriate color coding can significantly improve detection of targets among a few objects (typically less than ten). However, in spite of all the claims of the advantages of pseudo-color scales, to date, no study has demonstrated the superiority of any color scale—as can be objectively measured by improved performance—for continuous data, or data with large number of distinct values.

Todd-Pokropek claims the superiority of the gray scale over the Heated-Object scale for detection of small abnormalities in a reasonably complicated background [Tod83]. Levkowitz and Herman have also demonstrated the superiority of a linearized gray scale over their own Linearized Optimized Color scale (LOCS) and over the Heated-Object scale in a three-scale comparison study of target detection in a noisy background [LH92].

As Schuchard points out, this may be due to the fact that no large attention has been given to the development of color scales based on perceptual properties [Sch90b].

This chapter partially addresses this issue by presenting a method and an algorithm for the derivation of color scales, such that their perceptual properties—in particular, the perceptual steps between colors along the scale—can be controlled by the scale designer. This approach has been used for the design of the Linearized Optimized Color scale (LOCS). The results are demonstrated on the LOCS and the linearized gray scale, see Chapter 7; both are also compared to the Heated-Object scale.

Chapter 7 presents the main concepts and issues behind color scales. We use these here. Section 8.2 discusses some issues fundamental to the adjustment of perceptual steps along a color scale. Section 8.3 summarizes the main approaches to the linearization of color scales, and presents an algorithm we have developed.

8.2 ADJUSTING PERCEPTUAL STEPS OF COLOR SCALES

8.2.1 Numerical vs. perceptual steps

Most color scales are developed mostly based on the capabilities of the display device on which they are to be used. This usually means specifying the colors by their RGB primary values. In the worst case these are specified as digital-to-analog converter (DAC) values; those are totally meaningless once the display device changes. Even if the RGB values are calibrated—making them compatible across devices—their spacing does not correspond to perceptual steps.

8.2.2 Uniformity

One of the most desirable perceptual properties of a color scale is its ability to convey equally spaced numerical steps as equally spaced perceptual steps. Linearization of a color scale (also referred to as *equalization*) is a process in which additional colors are added to the scale—creating an extended, super-sampled scale—in such a way that the perceived distances between adjacent colors of the extended scale are as uniform as possible. This linearization can only be approximate, since it is based on estimating JND steps along the scale; these depend on the JNDs estimating metric, the observer, the display device, the illumination, and other conditions [PZJ82].

8.3 LINEARIZATION OF COLOR SCALES

Pizer [Piz81a, Piz81b, PZJ82] and Robertson [Rob85] describe algorithms to linearize color scales.

Pizer's linearization algorithm [Piz81b] works under the assumption that the scale at hand can be characterized by piecewise linear interpolations between few measured points (typically eight to ten points out of a 256-color scale). The algorithm is shown to perform very well on gray scales. The author reports that the algorithm has also performed nicely on a Heated-Object scale; performance of the algorithm on the Magenta and Rainbow scales are reported to have been less reliable, though the author attributes that to the fact that the necessary experiments have been carried out by less experienced students [PZJ82]. The author further states ([Piz81b]) that "[i]n each case the linearization has made an important difference not only in the appearance of an intensity wedge but also in the diagnostic usefulness of CT scans..." No details are provided on the reported differences in CT scan diagnostic usefulness. The strength of Pizer's approach, its being based on observer performance studies (which makes it potentially more accurate than other approaches) also presents a potential problem, in that its accuracy depends on the experience and competence of the implementor, and on the essential repeat and readjustment for each different observer.

Robertson [Rob85] proposes an algorithm that generates scales with m colors, where m is larger than n, the number of desired colors, and then re-scales to achieve the uniform spacing. The colors are generated in some Lightness, Hue, Saturation (LHS) space, and then converted by approximate rescaling to the CIELUV uniform color space (UCS), [CIE78].

8.3.1 Our approach

Our own algorithm, *LOS*—Linearize-Optimized-Scales—follows Robertson's approach. The main idea is to generate from the original scale a super-sampled, "long" scale (e.g., via interpolation) and to scan it for a subset of colors that are (almost) equally spaced perceptually. The main difference between our approach and Robertson's is that, while we specify our super-sampling phase in RGB coordinates as an example, we do not commit ourselves to any source color space, whereas, it appears like Robertson is committed to the Lightness, Hue, Saturation (LHS) color space as the source space.

Similarly—and perhaps more important—we do not commit ourselves to any particular uniform color space. The main reason for that is, that we have developed doubts about the global uniformity of the CIELUV color space.

8.3.2 Algorithm LOS

The input to *LOS* is N^o, the number of colors required of the output scale, and an input color scale with N^i color vectors, $\mathbf{c}_1^i, \ldots, \mathbf{c}_{N^i}^i$, whose component (primary) values are (e.g., in RGB space) $(0,0,0) \leq (r,g,b) \leq (M^i, M^i, M^i)$, $(N^i = M^i + 1)$ such that the lightness of colors in the scale—computed by the lightness function $\ell(\mathbf{c}) = (r+g+b)/3$—is $\ell(\mathbf{c}_{n^i}^i) = n^i - 1$ for all $1 \leq n^i \leq N^i$.

The output of *LOS* is a linearized scale with N^o colors, $\mathbf{c}_1^o, \ldots, \mathbf{c}_{N^o}^o$, whose primary values are $(0,0,0) \leq (r,g,b) \leq (M^o, M^o, M^o)$, $(N^o = M^o + 1)$ such

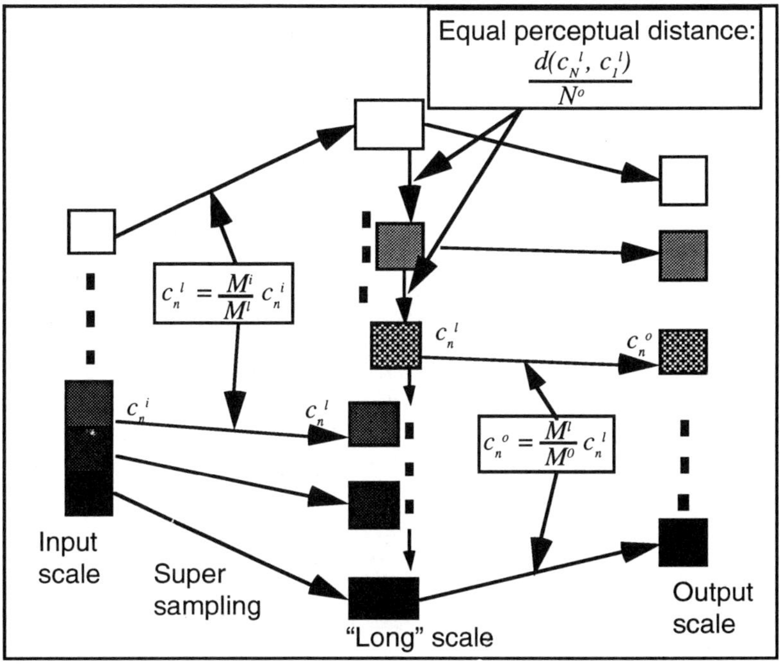

Figure 8.1 Illustration of the linearization of optimized color scales. See Algorithm *LOS*.

that the distance $d(\mathbf{c}^o_{N^o}, \mathbf{c}^o_1)$ between the first and last colors in the scale is divided (approximately) equally among all the color intervals, where the distance $d(\mathbf{c}, \mathbf{c}')$ between two colors $\mathbf{c}$ and $\mathbf{c}'$ is defined as the Euclidean distance between the two colors in the selected uniform color space. For example, in spite of our reservations about the global uniformity of CIELUV, for lack of a better choice, our implementation was actually done in CIELUV. Thus, in our specific implementation, $d(\mathbf{c}, \mathbf{c}') = [(L^*(\mathbf{c}) - L^*(\mathbf{c}'))^2 + (u^*(\mathbf{c}) - u^*(\mathbf{c}'))^2 + (v^*(\mathbf{c}) - v^*(\mathbf{c}'))^2]^{1/2}$.

LOS steps are (see Figure 8.1 for an illustration):

1. **Super sampling:**

   (a) **"Stretch" the input scale:** We compute from the input scale a new, longer scale with N^l colors whose values are (e.g., in RGB space) $(0,0,0) \leq (r,g,b) \leq (M^l, M^l, M^l)$, as follows. The maximum primary value M^l is chosen such that the scale features enough super sampling to provide sufficient resolution for the subsequent scanning for equally spaced colors in the scale (perceptual spacing is measured in some uniform color space, e.g., CIELUV). Then, the length N^l of the scale is set to be $N^l = 3 \cdot (M^l + 1)$ to enable storing colors in the order of their lightness (including colors with non-integer lightness) by using the value 3ℓ as an index to the location of a color in the scale. For each color $\mathbf{c}^i_{n^i}$ in the input scale, a color $\mathbf{c}^l_{n^l} = \frac{M^l}{M^i} \cdot \mathbf{c}^i_{n^i}$ in the long scale is computed whose location in the scale is $n^l = 3\ell(\mathbf{c}^l_{n^l})$, three times its lightness.

   (b) **Insert missing colors:** Stretching the original scale into the long scale leaves "empty color slots" between every two previously-adjacent colors (in the input scales). We now generate as many additional colors as necessary via interpolation or other techniques. (For example, for the linearization of our optimized color scale, OCS, to generate the linearized version, LOCS, we applied our optimized color scales algorithm to generate between each pair of previously-adjacent colors a "mini optimized scale," i.e., a scale that fulfills the optimization conditions required from the original optimized scale.)

2. **Scanning for equally spaced colors in UCS:** We scan the complete long scale resulting from steps 1(a) and 1(b) for a sequence of N^o colors that are equally distant from their adjacent colors and such that the sum of their distances equals the total distance $d(\mathbf{c}^l_{N^l}, \mathbf{c}^l_1)$ from the first color to the last one in the scale. An alternative approach to minimizing $|d(\mathbf{c}^l_{n^l_t}, \mathbf{c}^l_{n^l_b}) - j|$ in Algorithm LOS—where $\mathbf{c}^l_{n^l_b}$ is the bottom color in a step, $\mathbf{c}^l_{n^l_t}$ is the top color in a step, and j is the desired interval in UCS between two adjacent colors in the output scale—is to minimize $|d(\mathbf{c}^l_{N^l}, \mathbf{c}^l_{n^l_b}) - (N^o - n^o) \cdot j|$, where $N^o - n^o$ is the number of steps from the last color in the long scale, $\mathbf{c}^l_{N^l}$, to the current color n^o. This has the advantage of avoiding the cumulative error that can be introduced by the method described previously. We keep the algorithm as described for the sake of consistency with the computations we have conducted.

3. **"Shrink" the scanned scale:** For each color $\mathbf{c}^l_{n^l}$ in the long scale that is selected in the scanning process, we compute a color $\mathbf{c}^o_{n^o} = \frac{M^o}{M^l} \cdot \mathbf{c}^l_{n^l}$ in the output scale whose (r, g, b) values are $(0,0,0) \leq (r,g,b) \leq (M^o, M^o, M^o)$.

Algorithm *LOS*

Algorithm *LOS*
Input:

$\mathbf{c}^i_1, \ldots, \mathbf{c}^i_{N^i}$**:** A color scale of N^i colors whose primary values are $(0,0,0) \leq (r,g,b) \leq (M^i, M^i, M^i)$, $(N^i = M^i + 1)$ such that $\ell(\mathbf{c}^i_{n^i}) = (n^i - 1)$ for all $1 \leq n^i \leq N^i$.

N^o**:** The length of the output scale.

Output: $\mathbf{c}^o_1, \ldots, \mathbf{c}^o_{N^o}$, a linearized optimal scale of N^o colors whose primary values are $(0,0,0) \leq (r,g,b) \leq (M^o, M^o, M^o)$, $(N^o = M^o + 1)$ such that the distance $d(c_{N^o}, \mathbf{c}^o_1)$ between the first color and the last color is divided (approximately) equally among all the color intervals.

Auxiliary Variables:

$\mathbf{c}^l_1, \ldots, \mathbf{c}^l_{N^l}$**:** A long scale of N^l colors whose primary values are $(0,0,0) \leq (r,g,b) \leq (M^l, M^l, M^l)$, and $N^l = 3 \cdot (M^l + 1)$ (see explanation in text).

c**:** A temporary color variable.

n^l_b, n^l_t**:** Indices to the b(ottom) and t(op) colors of a segment in the long scale that is determined by two adjacent colors $\mathbf{c}^i_{n^i}, \mathbf{c}^i_{n^i+1}$ in the input scale (also used in the scanning of colors for the output scale).

j**:** The desired interval in UCS space between two adjacent colors in the output scale.

begin
    $\mathbf{c}^l_1 := \mathbf{c}^i_1$;
    $n^l_b := 1$;
    for $n^i := 1$ to $N^i - 1$ **do begin**   (stretch)
        $c := \frac{M^l}{M^i} \cdot \mathbf{c}^i_{n^i+1}$;
        $n^l_t := r(c) + g(c) + b(c)$;
        $\mathbf{c}^l_{n^l_t} := c$;
        INSERT-MISSING-COLORS;
                (e.g., for optimal scales: *OPTIMAL-SCALES*$(\mathbf{c}^l_{n^l_b} \rightarrow \mathbf{c}^l_{n^l_t})$)

$n_b^l := n_t^l;$
endfor;
$j := d(\mathbf{c}_{N^l}^l, \mathbf{c}_1^l)/(N^o - 1);$
$\mathbf{c}_{N^o}^o := \mathbf{c}_{N^l}^l \cdot \frac{M^o}{M^l};$
$n_t^l := N^l;$
$n_b^l := n_t^l - 1;$
for $n^o := N^o$ down to 1 **do begin**
repeat
$n_b^l := n_b^l - 1;$
until $|d(\mathbf{c}_{n_t^l}^l, \mathbf{c}_{n_b^l}^l) - j|$ is minimum;
$\mathbf{c}_{n^o}^o := \mathbf{c}_{n_b^l}^l \cdot \frac{M^o}{M^l};$
$n_t^l := n_b^l;$
endfor;
end; {*LOS*}

We ran Algorithm *LOS* with a goal to produce a linearized gray scale (LinGray) and a linearized optimized color scale (LOCS) that would be comparable to the optimized color scale we had generated using Algorithm *OPTIMAL-SCALES* [LH92]. In both cases we started with an input scale with $N^i = 64$ and used $M^l = 767$ (the length of the input scale was chosen such that it would be the longest possible to compute within a reasonable amount of time; the longer the input scale, the more accurate it will be). Thus, $N^l = 3 \cdot (M^l + 1) = 2304$. Using Algorithm *LOS*, we produced linearized scales with $N^o = 256$. The top of Figure 8.2 compares the display of a 0–255 wedge, where black is 0, white is 255, using (left to right) the gray scale; the linearized version, LinGray; the original optimized color scale, OCS; the linearized optimized color scale, LOCS; and the linearized Heated-Object scale. The bottom of Figure 8.2 compares brain slices displayed using (left to right) the linearized version of the gray scale, LinGray; the linearized optimized color scale, LOCS; and the linearized Heated-Object scale.

It is worth noting that the example pictures in the figures have passed several steps of reproduction, which have degraded their quality accuracy significantly. They should, thus, only be used for qualitative demonstration.

8.4 SUMMARY AND NOTES

We have described the concept of color scales for representation of data in general, and perceptual properties that can make such representations more

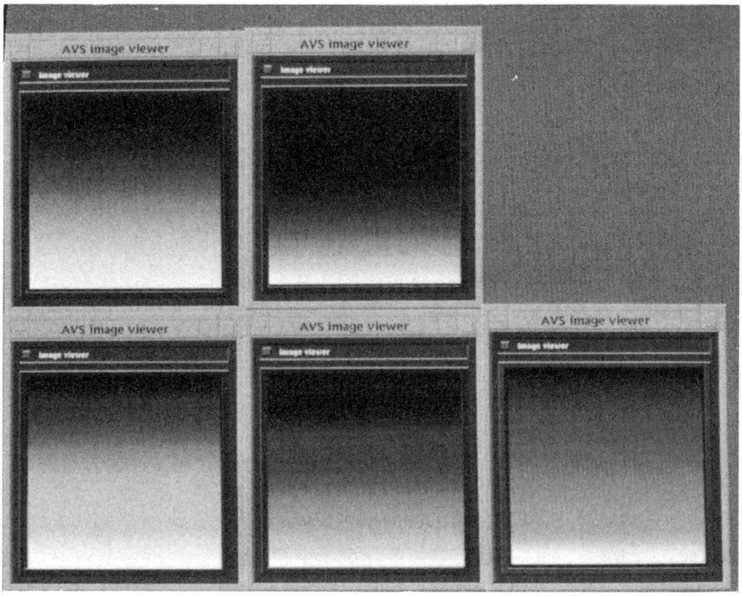

Figure 8.2 Scale comparison. A 0–255 wedge, where black is 0, white is 255, using the gray scale (top left); the linearized version, LinGray (top middle); the original optimized color scale, OCS (bottom left); the linearized optimized color scale, LOCS (bottom middle); and the linearized Heated-Object scale (bottom right). See Figure 7.9 for brain slices displayed using the linearized version of the gray scale, LinGray; the linearized optimized color scale, LOCS; and the linearized Heated-Object scale. (See Web site for color version.)

accurate, and thus more effective. We have discussed one approach, the linearization of color scales. We have described briefly two previous algorithms for linearization. We have described our own algorithm for linearization, and have shown two examples of its results: LinGray, the linearized gray scale, and LOCS, the linearized optimized color scale.

The main limitation of the results of the approach presented here is the use of the CIELUV Uniform Color Space as a measure of JNDs. The CIELUV space is considered to be reliable as a uniform space and as a measure of JNDs only within small localized neighborhoods in color space. Our approach requires the consideration of JNDs along the entire scale. Such broad range might introduce perceptual differences where equal numerical JND steps have been obtained. This would compromise the linearization process, and thus the uniformity of the resulting scale.

Further work is needed to compare these results with results obtained using other JND measures, which this approach can support. Additionally, observer performance studies are necessary to verify the uniformity of the obtained scales, and to compare them to their non-linearized counterparts, as measured by the performance of observers using the two classes of scales.

As mentioned in our Summary and Notes to Chapter 7, the linearization process can be incorporated as one of a number of techniques in the post-processing module.

9

INTEGRATED VISUALIZATION OF MULTIPARAMETER DISTRIBUTIONS

Multiparameter data collection is becoming increasingly common in many scientific disciplines. The data can be spatially coherent, such as medical images, satellite images, or geological data for oil exploration; or it can be non-coherent, such as population or census data. But it usually consists of a very large number of some higher dimensional vectors. In many such cases, a critical need exists to analyze the parameters not only individually—which is a relatively easy task for a human viewer, but also relative to each other—which can be extremely complex. The need exists for integrated displays to enable the scientist to see the combined information, some of which may not be adequately perceived, or may not be perceived at all when viewed individually. In order for such displays to succeed, they must exploit the visual perceptual capabilities as much as possible.

In this chapter we discuss the general problem, and describe the use of the GLHS family to address it.

In Chapter 10 we describe a more general approach, using *color icons*.

9.1 INTRODUCTION AND BACKGROUND

Visual analysis is very effective in situations where the patterns of potential interest are directly visible in *single* images. However, as multiparameter data collection—i.e., *multiple* images of the same scene obtained, each representing a different parameter—is becoming increasingly common in many scientific disciplines, there is a growing demand for analysis of such collections of images.

The scientist analyzing multiparameter image data is faced with several images representing different parameters, each of which contributes to the process of analysis. Detection and classification of objects and patterns in the various images require careful analysis not only of the separate parameters individually, but also of relationships among corresponding regions across them. Scanning back and forth among the separate parameter images positioned side-by-side can be adequate only for the analysis of crude aspects of relationship; more subtle changes in the shape or internal structure of corresponding regions may be very difficult to analyze. Moreover, in some cases, structures of interest may not be visible clearly, or at all, in any one of the separate parameters. They might, however, be visible in some appropriately integrated display. The need exists for such integrated displays to enable the scientist to see the combined information. In order for such displays to succeed, they must exploit the visual perceptual capabilities as much as possible. In particular, the image data have to be made directly visible in a single image.

The advantages of human visual analysis over machine analysis and the weaknesses of side-by-side displays have been discussed in various publications; for a summary see [EGL+95b, LP90, Lev91, PLS90, PG88].

As modern computer graphics and display technology provide a basis for many promising ways to integrate multiparameter images, two visually-based[1] techniques are becoming more and more recognized as powerful means to achieve that goal:

1. *Color integration*, which exploits humans' powerful color perception has the ability to integrate images, though limited to direct integration of no more than three parameters [Lev88, 35, LP90, PLS90].

   Very little is known about the absolute or relative merits of alternative color integration models [LP90, PLS90]. This is a very needed research, which has not been adequately supported so far.

2. *Iconographic integration*, which can employ color, shape, and texture perception, subsumes color integration. It is more general, and it does not suffer the limitation on the number of parameters that can be fused [Lev91, LP90, Pic70, PG88, PLS90]. In Section 9.2 and in Chapter 10 we describe this approach in some more detail.

[1] *Sonification* ([GS90]), a non-visual approach, which has shown tremendous potential in conjunction with visually-based techniques, is beyond the scope of this book.

9.1.1 Potential gains from integration

Levkowitz and Pickett identify three types of potential gains from integration [Lev91, LSPD90].

Type I: Gains in contrast and conspicuousness of structures.

Type II: Gains in the ability to classify structure types and in the ability to associate spatially disjunct regions of structures of the same type.

Type III: Gains from making structures that are not visible in any single parameter visible by combining parameters.

We now describe briefly the main approaches to visual integration; detailed descriptions can be found in the cited literature.

9.1.2 Color integration

In a color integrated display, a pixel's tristimulus values are controlled by the values of corresponding locations in (up to) three input parameter distributions. Typically, the different parameters are mapped to the different coordinates of a color model [Lev88, LH93]. Some desirable properties that would make a color model appropriate for this purpose are:

Perceptually orthogonal axes: Variations in one axis should not cause perceived variations in any other axis that is perceptually orthogonal to the axis along which variations occur. (This property is sometimes referred to as *separability.*) This is of particular importance if the ability to discern the three coordinates of a presented color independently of each other (sometimes referred to as *unmixing*) is desirable.

Uniformity: Uniform changes in parameter values along the axes of the color model should be perceived as uniform changes in the colors. This is important for accurate quantification.

Meaningful axes: The relationships between perceived colors and the values that define those colors should be clear and meaningful.

Ease of implementation: The transformations of color coordinates between the coding model and the display model (RGB) should be simple and efficient.

The easiest model to implement is the RGB model used in display monitors. The distributions of the different input parameters control directly the R, G, and B values fed to the CRT guns. No transformation computations are necessary. Also, the axes are meaningful to viewers with normal color vision. However, the model does not provide perceptual orthogonality or uniformity. Thus, images represented using this model might not provide good visualization of several parameters.

Models based on more perceptual primaries, such as the lightness, hue, and saturation (LHS) models, provide some perceptual orthogonality under certain conditions and range restrictions. They also provide meaningful axes and relative ease of implementation, though the most commonly used ones up to now, lack uniformity.

The Generalized Lightness, Hue, and Saturation (GLHS) family of models, described in Section 3.4 provides a parametrized implementation of the previous models as well as many other ones. As discussed in Chapter 4, among the new models in the GLHS family there might be ones that may provide better uniformity then the traditional LHS models. Also, the ability to change model representations might enhance the perception of certain features in the represented data.

9.1.3 Geometric integration

Geometric integration, originally proposed by Pickett [Pic70], has been developed at the University of Massachusetts Lowell for visualization of large multiparametric databases. It is a powerful approach to integration, in particular where the number of parameters exceeds three, which is the upper limit for color integration. In this technique shape perception is utilized to sense local combinations of data, while texture perception provides a sense of global spatial distribution of those combinations.

It has been well demonstrated that human observers effectively discriminate visual textures, and that variations in texture serve as important sources of information for space perception and locomotion [Gib50] and for the detection and recognition of objects [Pic70]. Psychophysical studies conducted to evaluate and model texture perception indicate a strong potential of the use of visual texture displays for exploratory data analysis. Though much more understanding is still needed to fully exploit this capability, there are some strong indications of the kinds of codings that might prove useful [Enn90, Enn91, LP90, Pic70, PG88, PLS90].

For example, research on *preattentive* processing of visual elements, indicates the ability to sense shapes or patterns as different without having to focus attention on the specific characteristics that make them different. Various studies [BPR82, Enn88, TG88] document the kinds of differences among elements that are preattentively discriminable. Variations of certain attributes of color, shape, and motion of elements lead to preattentive discrimination. Variations of such attributes, if brought under data control in texture displays, could improve the discrimination of various objects and structures in the data.

9.2 VISUALLY-BASED INTEGRATION TECHNIQUES: THE ICONOGRAPHIC APPROACH

The iconographic approach provides a general structure for the specification and creation of integrated displays of multiparameter images [Lev91, LP90, PG88, PLS90]. The basic primitive of this approach is the *icon*, which is a generalization of the standard notion of a pixel to higher dimensions having multiple perceivable features and attributes. It enables the representation of multiple measures at each spatial location by controlling attributes of the icon's perceivable features.

Gray-scale pixels in classical representations of images are the most degenerate form of an icon—a 1×1 patch with one attribute of variation, its intensity. Color coding of the 1×1 pixel patch provides the simplest extension, which is introducing three attributes of variation, its tristimulus values such as lightness, hue, and saturation. By expanding the area of the icon, albeit at potentially significant costs in spatial band-width, the icon may acquire many additional visual features and attributes, all of which are data-controllable. Additional perceivable features include shape and other aspects of visible geometry as well as the color of regions. Other potential extensions increase the dimensionality, either spatially—by adding stereoscopic depth, or temporally—by adding dynamic properties.

9.3 COLOR INTEGRATED DISPLAYS USING THE GLHS FAMILY OF COLOR MODELS

One potential use of the GLHS family of color models is for integration of multiparameter image data into a single display. This is becoming more desirable in several applications; [35, LP90, PLS90]. In color-based integrated displays, images of the different parameters are coded by different coordinates of some color model. In order to make the interpretation of such integrated displays easy and efficient, a color model is required whose coordinates are perceptually orthogonal. In such a model it is easy to determine the values of each of the coordinates of a viewed color independently of the others. Additional properties that are desirable are uniformity, intuitive axes, and ease of implementation. Analysis of various color models (e.g., RGB, LHS, CIELUV, The Munsell Book of Color) *vis-a-vis* these properties shows that the most promising model would be some form of a perceptually-based model (such as GLHS, CIELUV, or the Munsell model) [SBCG84, SCB87]. The GLHS family of models provides a unified implementation of existing and new LHS models, some of which may prove to be better suited for this particular task.

MCMIDS—Multiple Color Model Image Display System was developed to provides the capability to encode multiparameter images using any of these models. Model selection is similar to the one in the *MCMTRANS* software described in Chapter 3. In addition, file mapping to color parameters is supported. This allows the comparison and exploration of the potential merits of different color models for coding multiparameter image data into integrated displays.

To illustrate the potential of the GLHS models for coding multiple parameters (and the differences among the various models) we mapped a scale on the vertical axis (0–255, bottom to top) to lightness and a scale on the horizontal axis (0–240, left to right) to hue, keeping saturation constant at the maximum value 1. Figure 9.1 shows the envelopes of the color solids of the same GLHS models used for Figure 3.9. Note the different locations of primaries and secondaries (mixtures of two primaries) in each model.[2]

[2] The perceived variations in hue along the vertical axis (where the hue is constant) are due to several causes, such as context, which can be reduced by viewing narrow vertical strips; the interaction between lightness and hue variations; and the degradation in color quality due to the various reproduction steps.

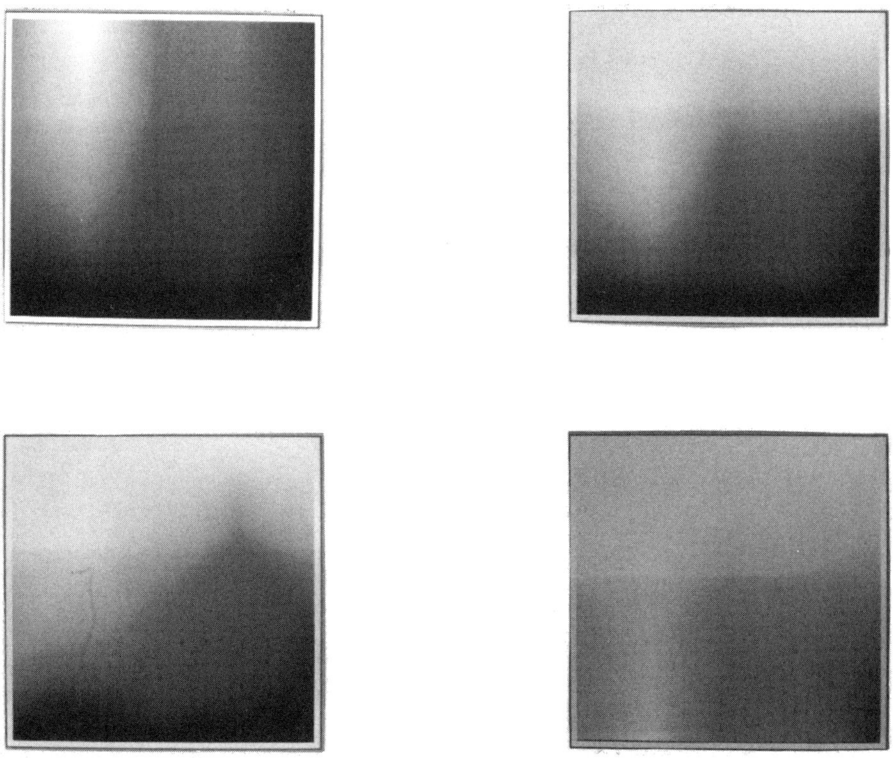

Figure 9.1 MCMIDS—Multiple Color Model Image Display System. Lightness: vertical axis (bottom: $\ell = 0$, top: $\ell = 255$). Hue: horizontal axis (left: $h = 0$, right: $h = 240$). Top left: $(w_{max}, w_{mid}, w_{min}) = (1,0,0)$ (HSV-hexcone); Top right: $(w_{max}, w_{mid}, w_{min}) = (0.5,0,0.5)$ (HLS-double-hexcone); Bottom left: $(w_{max}, w_{mid}, w_{min}) = (0.333,0.333,0.334)$ (LHS-triangle); Bottom right: $(w_{max}, w_{mid}, w_{min}) = (0.001,0.009,0.99)$. (See Web site for color version.)

9.4 SUMMARY AND NOTES

We have discussed the concept of integrated visualization of multiparameter distributions. We have examined some potential gains from such integration, and described the two main visual approaches: color and geometric integration.

Levkowitz et al. have shown potential improvements in the detection and classification of various tissues types in color integrated visualizations of combined CT and MR images of the abdomen [LSPD90]. In that limited study, radiologists rated characteristics of contrast and identifiability of different tissue types in various color integrated and gray-scale image configurations. For some contrast tasks, the best rated configuration was a color integration that used one of the GLHS models.

The concepts of color and geometric integration lead to the more general concept of iconographic integrated visualization. We introduced that concept. Chapter 10 explores the combination of color and iconographic integration in much more depth.

10

COLOR ICONS: MERGING COLOR AND TEXTURE PERCEPTION FOR INTEGRATED VISUALIZATION OF MULTIPLE PARAMETERS

In Chapter 9 we discussed the general problem of multiparameter integrated displays, and demonstrated color integration using the GLHS family of color models. In this Chapter[1] we introduce a more general approach, the color icon.

The color icon presented in this chapter, harnesses color and texture perception to create integrated displays of multiparameter distributions [EGL95a, Lev91].

We first describe the rational behind the development of the color icon, followed by the original color icon design. We then demonstrate the power of the technique with some examples using real and synthesized data, comparing it with several other proposed techniques.

After some experience with the color icon, it was redesigned. The new design increases the number of parameters that can be integrated. It also extends the icon from two dimensions to three dimensions (which increases the number of parameters that can be integrated even further). To facilitate comparison with other iconographics approaches, the color icon has been incorporated as one of several other icons within the NewExvis environment.[2] Finally, a parallel version was implemented on a proprietary multiprocessor architecture. We describe the new design; the main issues, considerations, and features of the parallel implementation; and a few application examples.

We discuss the nature of studies required to measure objectively and accurately the effectiveness of such displays.

---

[1] This chapter is adapted from [Lev91] and [EGL95a].

[2] NewExvis is the latest version of Exvis, an exploratory visualization environment developed at the Institute for Visualization and Perception Research at the University of Massachusetts Lowell [EGL+95b].

10.1 MERGING COLOR, SHAPE, AND TEXTURE PERCEPTION FOR INTEGRATION: THE ORIGINAL COLOR ICON

The color icon was designed to merge separable features by using color, shape, and texture perception to code multiple parameters into a single integrated image.

It can be shown that shape and color are perceptually *separable* features; this means that discrimination of variations in these two features are perceptually orthogonal. Lightness and size are also separable.[3] Separable features are considered by some to have the best potential as vehicles for successful coding of data, in particular for unmixing of parameters.

At the most general level, the color icon is an area on the screen, constructed of one-dimensional features (boundaries and area subdividers), two-dimensional features (areas defined by the subdividers), and their attributes (orientation, size, shape, and color). Each of those can be controlled globally (e.g., the shape of the icon is chosen to be rectangular for *all* icons), or can be mapped to be controlled by a data parameter (e.g., a subdivision point along a linear feature is controlled by a parameter value) [Lev91].

10.1.1 A prototype implementation

In the first implemented prototype of a system to visualize data using the color icon, the icon is defined over an $m \times n$ rectangle (where m and n are under user control), see Figure 10.1, [Lev91]. In one variant of the prototype, six linear features are used: two of the edges (one horizontal, one vertical), the two diagonals, and the two midlines. This provides coding of up to six parameters (in addition to the (x, y) spatial location of the icon). The other two edges are not used in order to maintain boundary consistency by avoiding conflicts with the neighboring spatial points. Those two edges could, however, be used for two additional parameters. The features subdivide the icon's area into several subareas. The colors of those subareas are interpolated from the colors assigned to the linear features. In another variant, the features used are the fields resulting from subdividing the icon's area.

[3] On the other hand, lightness, hue, and saturation, in particular at the extremes of the lightness range, are *integral*, meaning that variations in any one might cause perceived variations in the others.

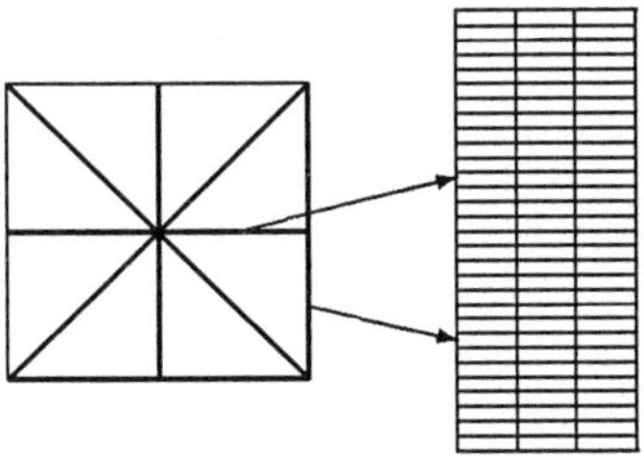

Figure 10.1 The color icon with six linear features: two edges (bottom horizontal, right vertical), the two diagonals, and the two midlines. A feature's color is determined by the value of the parameter associated with it, which points to a color scale. The colors of the inscribed areas are interpolated from the colors of the features.

Each feature is mapped to one data parameter. In the current prototype the attribute under data control is the color of the feature. Thus, the value of that parameter determines all three color coordinates[4] via a pointer to a color scale. The color pointed-to in the scale is assigned to the feature. Different parameters may share the same color scale or may point to individual color scales.

Figure 10.2 shows the picture of an enlarged color icon whose area is subdivided into six regions to support six parameters. Each parameter points to a separate color scale.

The first prototype system was implemented on a number of workstations in C, under X-windows using the OSF/Motif toolkit. It supported the loading, separate display, and individual color scale assignments for up to six parameters. Because a color-icon integrated image occupies a larger area on the screen than the input images, a region-of-interest selection capability allows displaying part of the image in the integrated display. In such a case, the corresponding regions in all input images are selected.

10.1.2 Other possibilities

The general definition of the color icon allows for many variations:

[4]This is only one approach to coding; we discuss other approaches later.

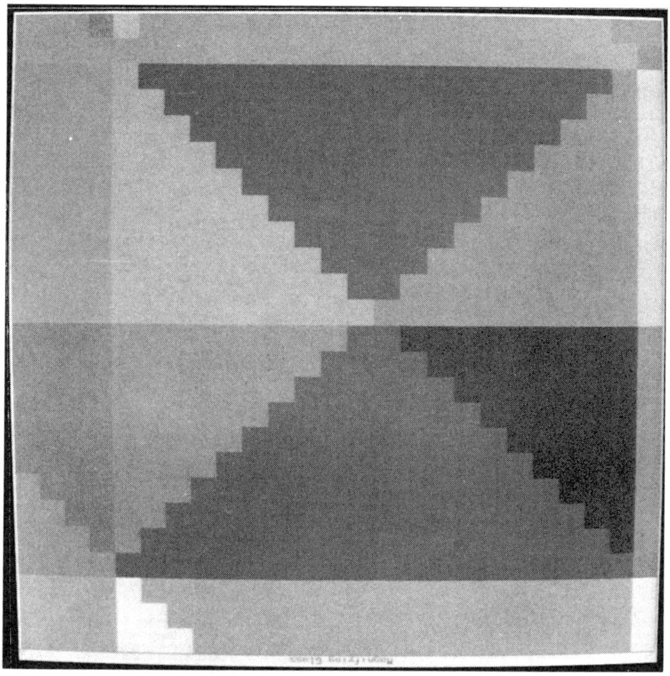

Figure 10.2 An enlarged color icon whose area is divided into six fields to support six parameters. Each parameter points to a separate color scale. (See Web site for color version.)

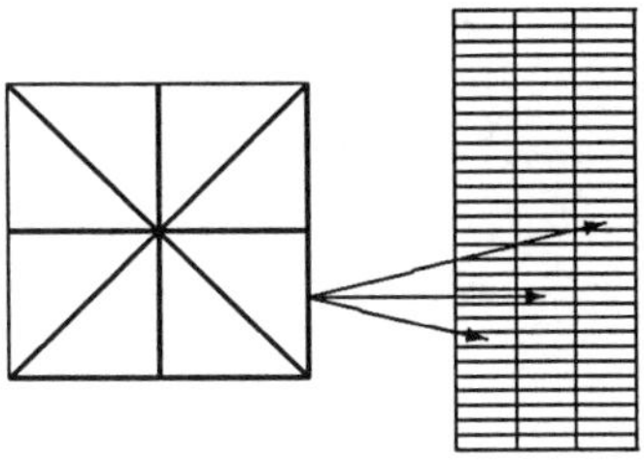

Figure 10.3 The number of parameters can be tripled by letting each parameter control only one of the three color coordinates of a feature's color (compare to Figure 10.1).

- The number of parameters to be coded can be tripled by allowing each parameter to control only one color coordinate (such as the lightness, hue, or saturation) of a feature. In such a case, three parameters are needed to control the color of a feature, as compared to the current prototype, where only one parameter controls the color of the feature via a pointer to an entry in a color scale, compare Figures 10.1 and 10.3.

- More than one parameter can be mapped to a linear feature by subdividing its length. Subdivision can be fixed globally (e.g., all linear features are subdivided in the middle) or can be data-controlled, where the point of subdivision is controlled by the value of a parameter.

- The contributions of the different parameters to the colors of the icon can be weighted. Weighted contributions can be achieved algebraically (by using a different interpolation function to compute the relative contributions) or geometrically (by changing the location of intersection points, the contributions of different parameters to the colors of the icon are given different weights). The two approaches can be combined.

10.2 ILLUSTRATIONS AND EXAMPLES

To illustrate the potential of iconographic displays as the ones described here, we present a set of synthesized images and a set of collected medical images.

Pairwise **negative** correlates	Pairwise **positive** correlates
$X_1 + X_2$	$X_4 + X_5$
$-X_1 + X_3$	$X_4 + X_6$
$-X_2 - X_3$	$X_5 + X_6$

Table 10.1 Pairwise negative and pairwise positive correlates generated for the synthesized data.

10.2.1 Synthesized images

In order to demonstrate the capabilities of different mechanisms to code multiple parameter images into integrated displays, we have developed a set of synthesized images that possess some known properties and relationships that are not available in real images. In particular, this synthesized data set has the property that structures and relationships inherent in the images are not perceivable when viewed separately, see Figure 10.5. These relationships become perceivable when integrated appropriately, see Figures 10.6–10.9. This illustrates our conjecture that in some cases intergrated displays may be the only effective way to make some relationships in the data perceivable, see also Visual Cues [KK82].

Image generation

To generate the images we first generated samples from the normally distributed random variables $X_1, \ldots, X_6$ with mean $\mu = 0$ and standard deviation $\gamma = 1$. Next, we generated the negations $-X_1, -X_2, -X_3$ and the pairwise negative and pairwise positive correlates as shown in Table 10.1. Finally, we generated the images as shown in Figure 10.4. The images are shown in Figure 10.5. Note that the top-bottom differences in the images are not perceivable at all in the gray scale representation. On the other hand, those difference are very clear in the color model codings shown in Figure 10.6 and in the color icon codings shown in Figures 10.7–10.9. This demonstrates the potential gains of Type III discussed in Section 9.1.1.

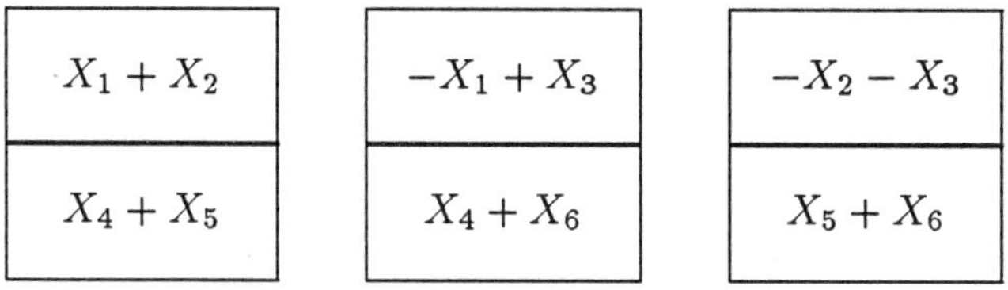

Figure 10.4 Generation of the three synthesized images.

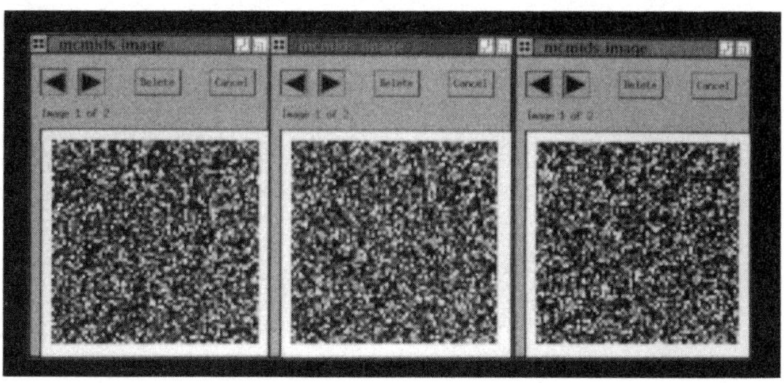

Figure 10.5 The three synthesized images displayed in gray scale. Note that the top-to-bottom differences in the data are not seen. (See also a version on the Web site.)

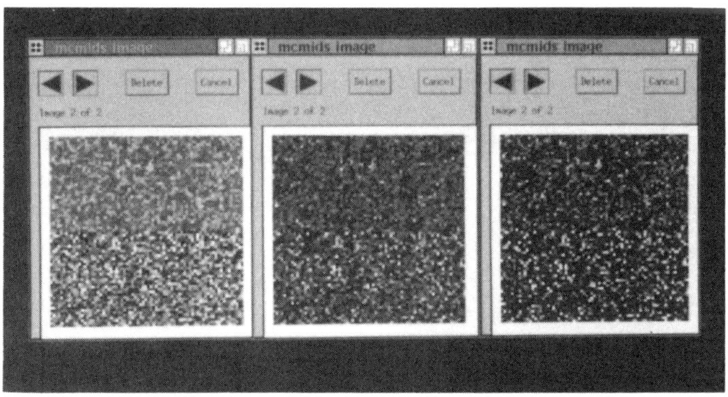

(a) (b) (c)

Figure 10.6 The three synthesized images displayed in three different color model codings: (a) RGB, (b) GLHS (1, 0, 0) [Lev88, LH93], (c) CIELUV. Note that the top-to-bottom difference in the data is very clear in all the color models. (See Web site for color version.)

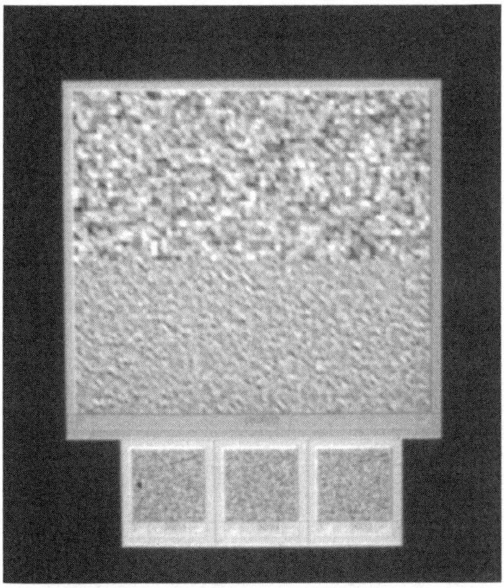

Figure 10.7 The three synthesized images displayed with the color icon. All three parameters displayed in the LOCS scale, [Lev88, LH92]. (See Web site for color version.)

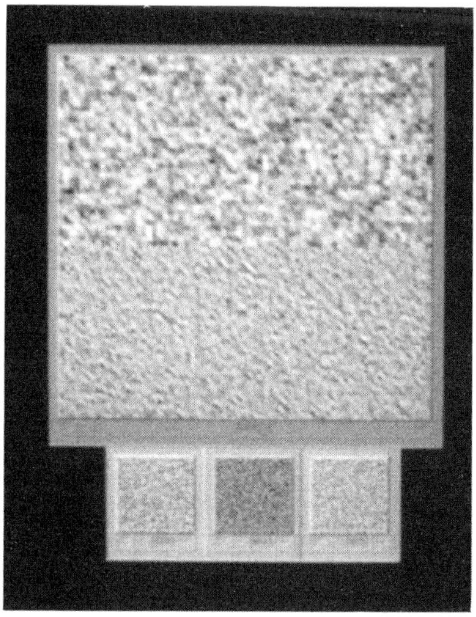

Figure 10.8 The three synthesized images displayed with the color icon. Each parameter displayed in a separate scale (Heated object, LOCS, magenta). (See Web site for color version.)

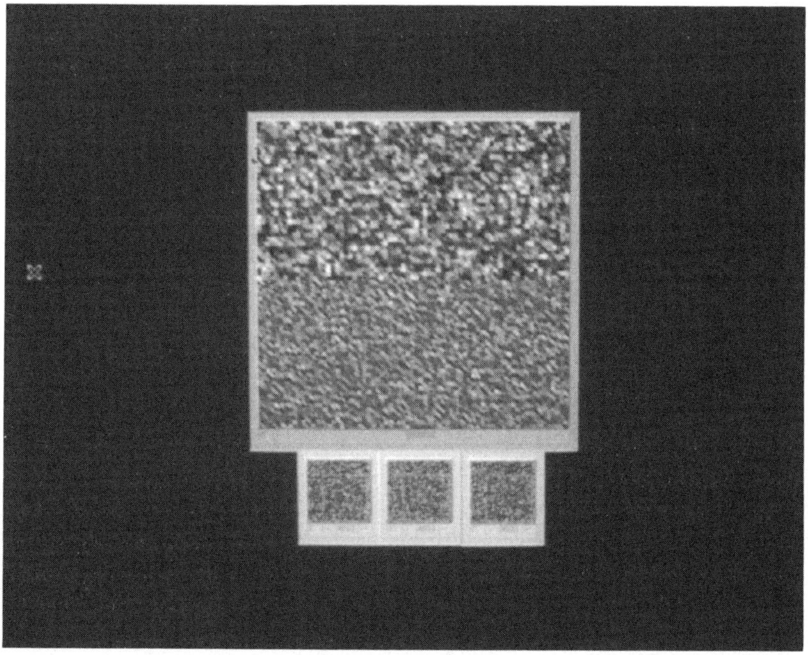

Figure 10.9 The three synthesized images displayed with the color icon. All parameters displayed in the gray scale. Note that the top-to-bottom difference in the data is seen clearly in all mappings. (See Web site for color version.)

10.2.2 Medical images

Figure 10.10 shows two MR images of the brain of a patient with a malignant glioma. The image in (a) is a so-called T1-weighted image. The image in (b) is a T2-weighted image. Both images demonstrate a lesion in the left frontal lobe extending into the temporal regions. The lesion is brighter than normal brain in both images. It is sharply marginated and slightly heterogeneous. We have generated integrated displays similar to the ones shown for the synthesized data shown in Figures 10.5–10.9. Figure 10.11 shows the two images coded with the same color models used in the illustrated synthesized data. Figures 10.12a and 10.12b show two color-icon-integrated images of the same brain section using the color icon. In Figure 10.12a, the image of Figure 10.10, left, is displayed using the heated-object scale [PZJ82] (left top). The image of Figure 10.10, right, is displayed using the Linearized Optimal Color Scale (LOCS) [Lev88, LH92] (left bottom). A region of interest has been selected; it is marked on both input images. This region of interest is displayed using a two-parameter color icon (right). Figure 10.12b was generated using the same process except that the mapping of color scales to the two input images has been swapped. Note that each integrated image emphasizes different features in the data.

10.3 SECOND GENERATION: NEW DESIGN AND IMPLEMENTATION

The original color icon discussed above ([Lev91]) introduced a linearly interpolated box icon. The color of each of the box's edges (four) is controlled, each, by a parameter value pointing to an entry in a color look-up table. Two diagonals and two mid-lines (horizontal and vertical), controlled in the same way, provide the potential for four additional parameters. Linear interpolation of colors provide the final color of each pixel within the icon.

After some experience with the color icon, it became clear that improvements could be made to the original design. This led to a new design, described here.

The new design changes the way colors are determined within the icon. Parameter values (given in database fields) are mapped to three parameters, referred to as "red," "green," and "blue," associated with each of the icon's attributes, called *limbs*. Note that the names of the parameters, "red," "green," and "blue," do not restrict them only to the RGB color model; the parameters are remapped to other color coordinates, depending upon the selected color model. The possible mappings are:

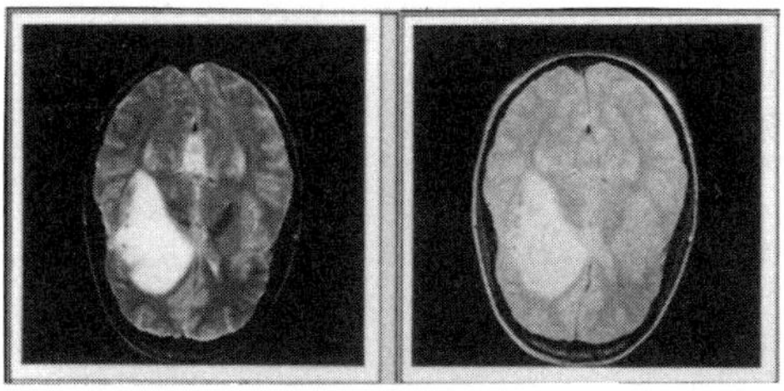

Figure 10.10 Left: A T1-weighted brain MR image of a patient with a malignant glioma. Right: A T2-weighted image of the same brain section. (See also version on Web site.)

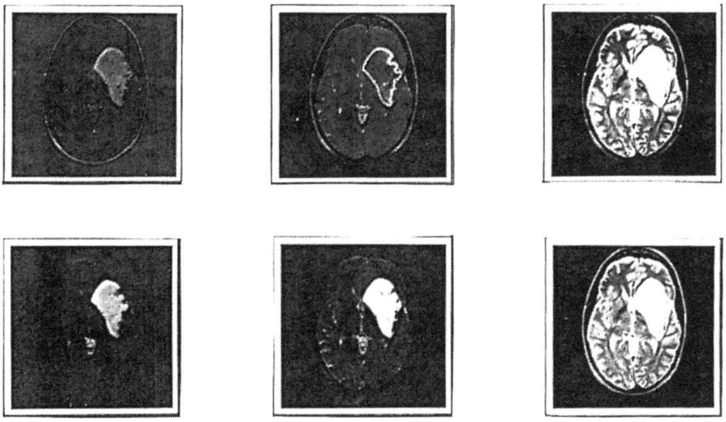

(a) (b) (c)

Figure 10.11 The two brain MR images of Figure 10.10 displayed in three different color model codings and two mappings with each model: (a) RGB: (Top) T1 → R, T2 → G, B; (Bottom) T2 → R, T1 → G, B; (b) GLHS (1, 0, 0): (Top) T1 → L, T2 → H, S; (Bottom) T2 → L, T1 → H, S; (c) CIELUV: (Top) T1 → L*, T2 → u*, v*; (Bottom) T2 → L*, T1 → u*, v*. (See page 193 and Web site for color versions.)

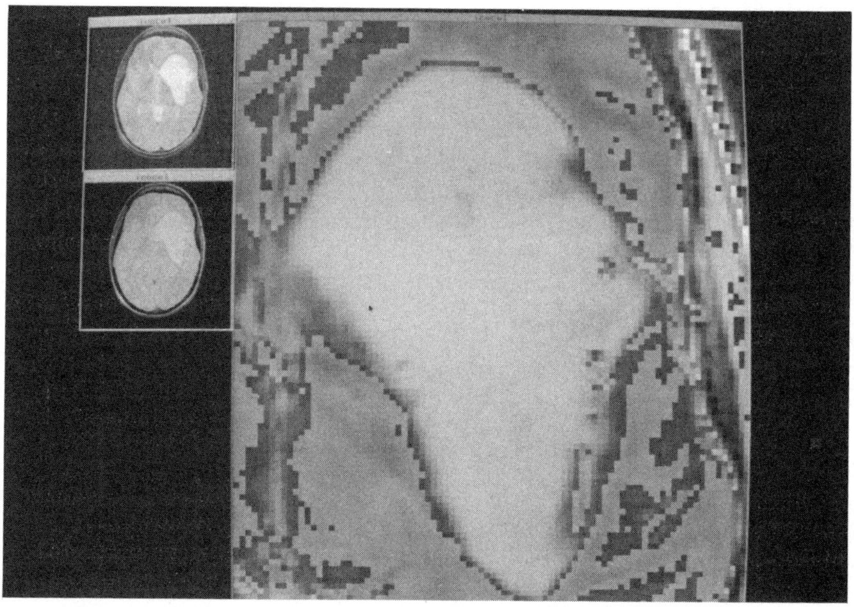

(a)

Figure 10.12a A color icon integrated image of the two brain sections. Two parameter-to-color-scale mappings are shown. T1 in LOCS, T2 in heated-object scale.

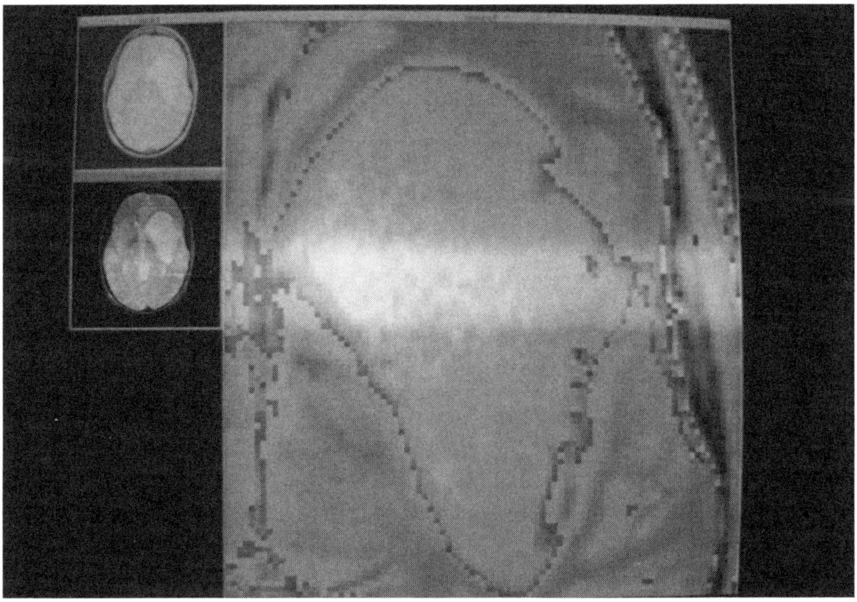

(b)

Figure 10.12b Another color icon integration of the same section; T1 in heated-object, T2 in LOCS. (See Web site for color versions.)

Parameter	Selected Color Model			
Label	RGB	GLHS	CIE	MUNSELL
"red"	red	Lightness	L*	Value
"green"	green	Hue	u*	Hue
"blue"	blue	Saturation	v*	Chroma

Thus the total number of parameters that can be mapped is three times the number of limbs; in our current example, with up to four limbs, we can map up to 12 parameters.

The current color icon configures its limb placement dynamically based on the number of parameters that are mapped. The limb configurations for one, two, three, and four limbs are shown in Figures 10.13, 10.14, 10.15, and 10.16, respectively. A label of $Limb_n$ on a corner means that the color of that particular corner of the icon is driven by the fields mapped to $Limb_n$. Colors within the icon are interpolated as described for each icon. (Arrows represent directions of interpolation.)

10.4 COLOR ICON SURFACES

We have further extended the color icon to three-dimensional surfaces. A color icon surface uses a quadrilateral mesh to combine up to 13 parameters in a single color image. The (x, y) coordinates of the four points of each quadrilateral are mapped the same way as the four points on the 2D color icon. In addition, a height parameter is mapped to the z coordinate of each of the four points. Currently, this height parameter is the same for all four points. This results in a flat, level surface for each color icon with all resulting normals pointing in the same direction. Allowing the z coordinate to be different for each point in the color icon would result in a total of up to 16 parameters being supported. This would also allow the normal vector for each color icon facet to vary, which may produce some interesting three-dimensional visual effects, especially if lighting is used.

10.5 PARALLEL IMPLEMENTATION

In [SGB91], Smith et al. describe a supercomputer implementation of the stick-figure icon [PG88], which showed the benefits of using a supercomputer

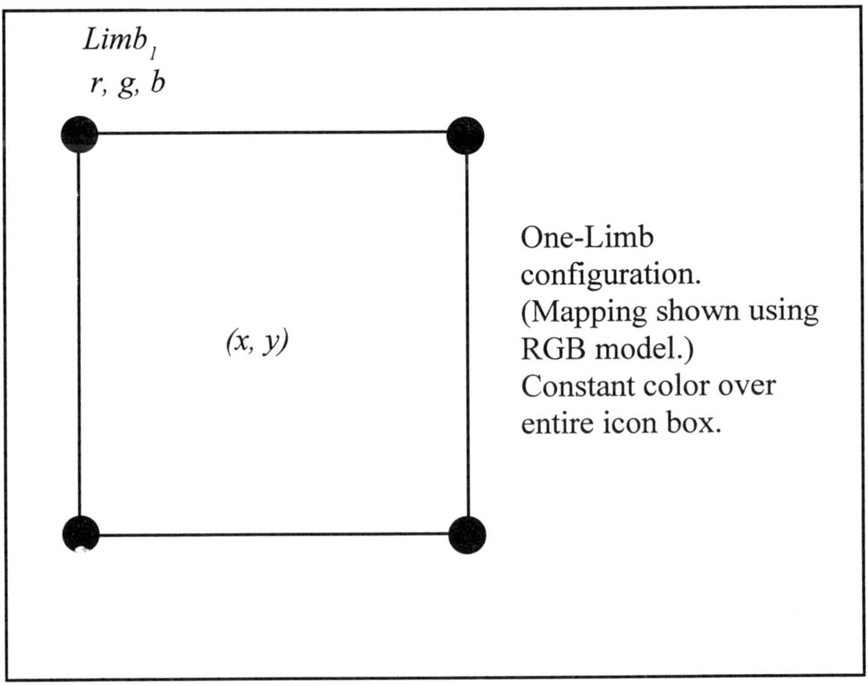

Figure 10.13 One-Limb Configuration.

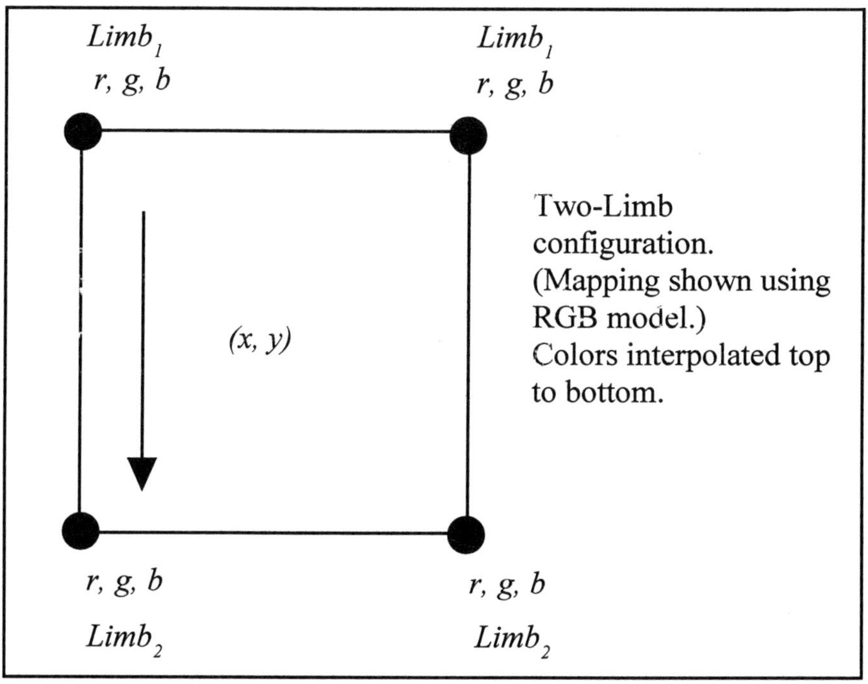

Figure 10.14 Two-Limb Configuration.

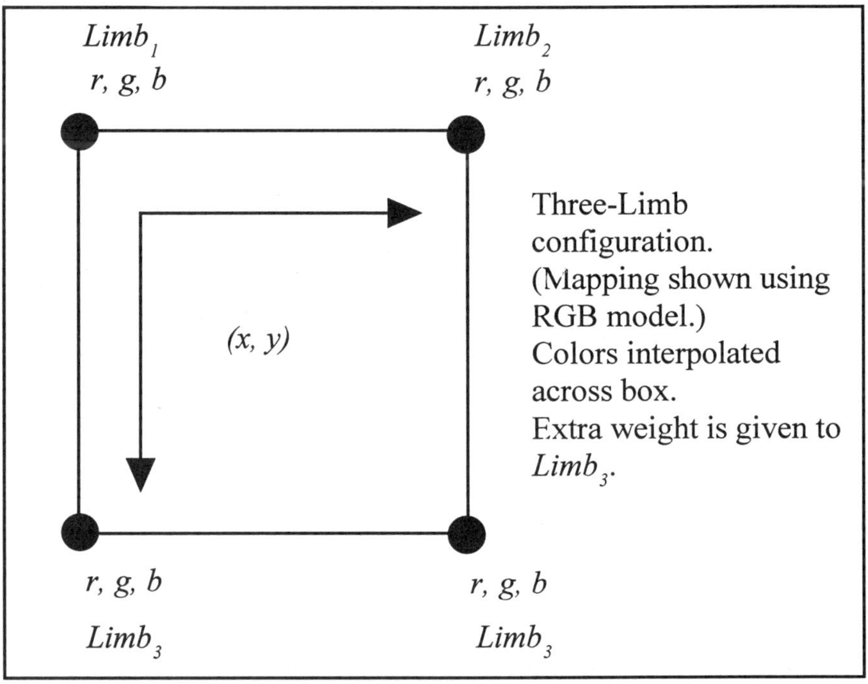

Figure 10.15 Three-Limb Configuration.

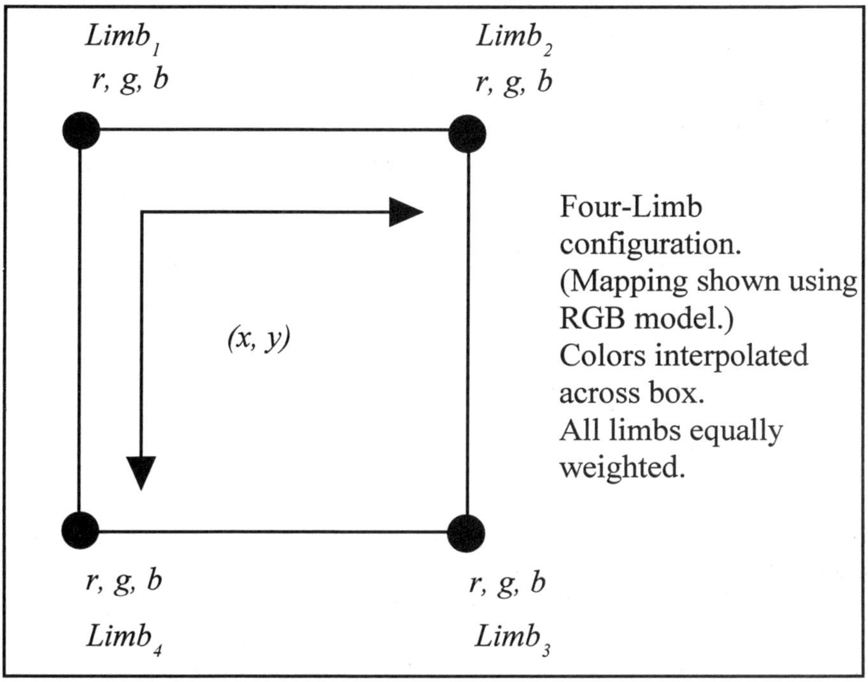

Figure 10.16 Four-Limb Configuration.

for data exploration using the stick-figure icon. Since the color icon is even more computationally expensive than the stick-figure icon we felt a supercomputer implementation of the color icon would be even more beneficial.

Under a grant from the Institute for Defense Analysis we have implemented a SIMD parallel version of the color icon renderer on a proprietary architecture. We discuss this implementation, its benefits, and performance.

Since the number of parameters and controls provided to the user in this application tends to be rather large, the number of possible combinations is extraordinary. One of the goals of this project was to provide a system that allows the user to explore the possibilities much more quickly than otherwise possible. We also aimed at providing a system that would allow the user to gain greater insight into how the parameters and controls affect the display; thus allowing the user to make more intelligent use of the controls.

10.5.1 Applicability to a super computer implementation

The serial implementation of the color icon generates a complete picture by rendering icons one at a time for all the points in a data set. Because each icon in a fully rendered data set can be calculated completely independently, there is the potential for extensive parallelization of the rendering routine.

The computational cost of the serial algorithm for rendering each icon varies not only with the color model selected, but also according to the data values given to the rendering routine. The reason is that the rendering algorithm depends on conditionals. Because each icon is generated from different input data, the serial version requires varying computational time for rendering each icon. By contrast, the SIMD paradigm forces all processors to execute in lock-step, and thus shows identical computation time for each icon generated. Every processor must step through each instruction in a conditional even if the instructions are not to be executed because the conditional evaluated to false. This behavior wastes cycles that could be spent on other tasks. The renderer is thus best suited for MIMD parallelization, which should offer a speedup over the SIMD implementation described here since a MIMD parallelization will not waste as many cycles. This analysis is based on the computation time required to generated icons, and ignores the issues of displaying the data, which are beyond the scope of this chapter.

10.5.2 The terasys system

The hardware provided to us for this project is the "Terasys" system developed by the Supercomputing Research Center [MN93, Swe93]. This system provides a bit-serial SIMD architecture in a linear array organization. The system we were provided with currently contains 4K processors, although up to 32K processors are supported. Each processor operates on a local 2K-bit memory, which is allocated to variables whose lengths may be specified. The Terasys also provides a parallel prefix network, which improves the performance of some important operations. A SUN Sparc II plays host to the Terasys hardware, which resides on the S-Bus. The performance of the S-Bus was found to be a limiting factor for real-time graphical displays.

10.5.3 User interface

The user interface as implemented in the serial version expects the displayed image to be static. Thus, the image is stored in a pixmap and the user must explicitly force the system to re-render the image when all desired changes have been made to the controls. This is satisfactory when considering the time required for the updates.

The interface had to be modified to update the display as the input parameters are changed rather than waiting for the user to force an update. Sliders are well suited for this purpose as they allow the user to scan over a range of values very quickly. For many applications sliders tend to be rather crude as they make it difficult to accurately select specific values for a parameter—which is generally desired for non-interactive updates—without switching input devices. An attempt was made to have the image be redrawn as the icon mappings are changed but this proved to be too slow to be usable, due to the need to reload too much data onto the Terasys. The triangular baricentric potentiometer used to select specific models in the GLHS (see Figures 3.9, 9.1) color model family was also modified to update the displayed image when its values are changed.

The most noticeable difference between the serial and parallel implementations is the method of handling the image window. In the serial version the image window is given a default size that can then be adjusted manually. This allows the user to make the window size larger, to display more data, or smaller, to speed up display. Currently, the system will automatically begin redrawing the image when the image window is resized, without user intervention. This can be annoying at times due to the amount of time required to redraw the image. Since it is rare to be able to display an entire database on screen at one time the region to be displayed is selected using a selection window that contains a scaled-down image of the original database as well as a selection rectangle. The selection rectangle designates the area to be displayed in the image window and is sized to show the area that will be encompassed by the image window at its current size. This selection window may then be placed to select the desired area for display.

In the parallel version the size of the image window is fixed. It depends on the number of processors, the size of each icon, and the algorithm used to render the image. The region selection window is nearly identical to the serial version except the display updates as the selection rectangle is dragged across the scaled down image.

10.5.4 Implementation

Implementing the parallel version of the color icon to achieve the desired goals required that we give up some of the generality of the serial routines. One of the goals of the serial implementation was to provide very general routines that could be reused to help make the generation of new icons easier. Much of this generality had to be given up in the parallel implementation to provide the performance needed for interactive updates.

An interesting issue, brought up by the need for good performance in the parallel implementation, was how the interpolation across the icon is done. In the serial version, interpolation is done before the data is converted from the current color space to the RGB color space. Thus, a complex filter must be passed over every pixel in an icon to convert the pixels into the RGB color space. While developing the parallel implementation we realized that better performance could be achieved if the four corners (limbs) of the icon were to be converted to the RGB color space and then interpolation was to be done directly on the remaining pixels without the need to convert them between color spaces. Although these two methods for interpolation provide slightly different results the method used in the serial implementation had been chosen arbitrarily. Consequently, we modified the parallel implementation to conform with the more efficient interpolation method in the hope that the two methods could then be compared and the most useful method be determined.

Currently, the parallel implementation generates an entire icon for each processor. This provides a reasonably sized image considering the machine we developed the parallel implementation on contains only 4K processors. If a machine with a larger number of processors is made available then each processor can generate less than a full icon—giving each processor less work to do.

10.5.5 Performance

The Terasys hardware proved to be capable of performing the calculations required for our system to generate pseudo real-time displays. The actual display of those images proved to be severely limiting and greatly decreased the frame rate. On our system the current implementation of the algorithm generates a 64×64 icon image (320×320 pixels). This image is calculated on the average in approximately 0.05 seconds. The actual display of the image adds about 0.15 seconds. Thus, while we could generate 20 frames per second without display we are limited to only five frames per second with display. Image calculation time can be improved with a different implementation of the algorithm, as discussed above.

Improving the display time is a much more complicated issue. Since both the Terasys hardware and the framebuffer are on the S-bus, a system-friendly display routine will require two passes over the S-bus. The first pass requires the data to be sent from the Terasys hardware to main memory, where it is then passed to X windows, which displays the image. Performance can be increased dramatically if some mechanism can be found to bypass X and have the data sent directly from the Terasys to the framebuffer. Another alternative is to incorporate a framebuffer directly onto the Terasys hardware. This would completely eliminate the need to transfer displays over the S-bus.

10.6 APPLICATIONS AND EXAMPLES

We now give several examples of applications where we have used the color icon.

10.6.1 Visible and infrared image fusion

The problem we set out to solve under a Litton/ITEK research grant was to devise a presentation technique that integrates a thermal (IR) image and a visible image into a single picture that incorporates as many of the unique features of both images as possible. The goal for visible/thermal sensor fusion is to provide a composite image that can be easily and intuitively exploited to rapidly extract the relevant information contained in both of the independent acquisitions of the scene. For this study we selected a pair of registered images that were collected nearly simultaneously. Two cameras were used to separately acquire the visible and infrared images.

Figure 10.17 shows the visible-band image (top) and the thermal image (bottom) acquired.

The scene shown in these images is a portion of Boston's Logan International Airport and a large industrial building nearby. Among the items of interest in the scene are:

- The smokestacks (we can see from the thermal image that only two of the four smokestacks are in use).
- The crane and associated machinery (the thermal image shows a "hot-spot" at the joint between the two sections of the crane).
- Details present in the visible-band image that are not evident in the thermal image (such as the windows in the large industrial building in the background).

Because thermal and visible sensors usually provide imagery of vastly different tonal presentations (thermal imagery often appears as a "negative" presentation when compared to a visible image) it is difficult to use simple image combination techniques to develop an optimum fusion technique. For example, simple image arithmetic often fails to incorporate subtle details in the dark regions of either image, and temporal interleaving of images can become very distracting because of the rapid alternation of negative- and positive-type images.

Smith and Scarf have previously shown that the iconographic approach can provide a successful fusion of the two images [SS92]. They used a two-limb stick-figure icon for the fusion [PG88].

Figure 10.18 shows our color icon fusion. The two active smokestacks are clearly differentiated from the two inactive ones and, at the same time, details such as the windows in the large industrial building are retained. The hot spot at the joint between the two sections of the crane stands out prominently. Thus, this inconographic presentation achieved the stated goals of the project.

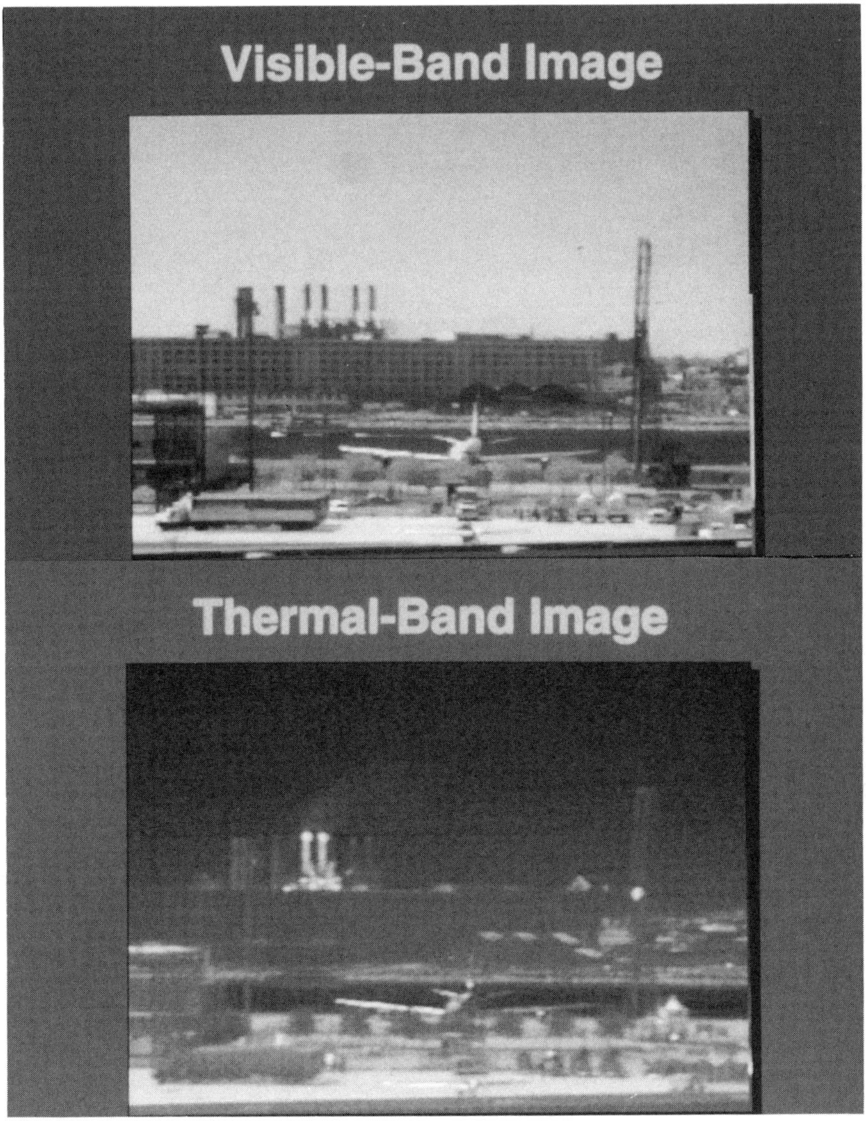

Figure 10.17 Original images used to implement the iconographic technique for sensor fusion. (See Web site for color version.)

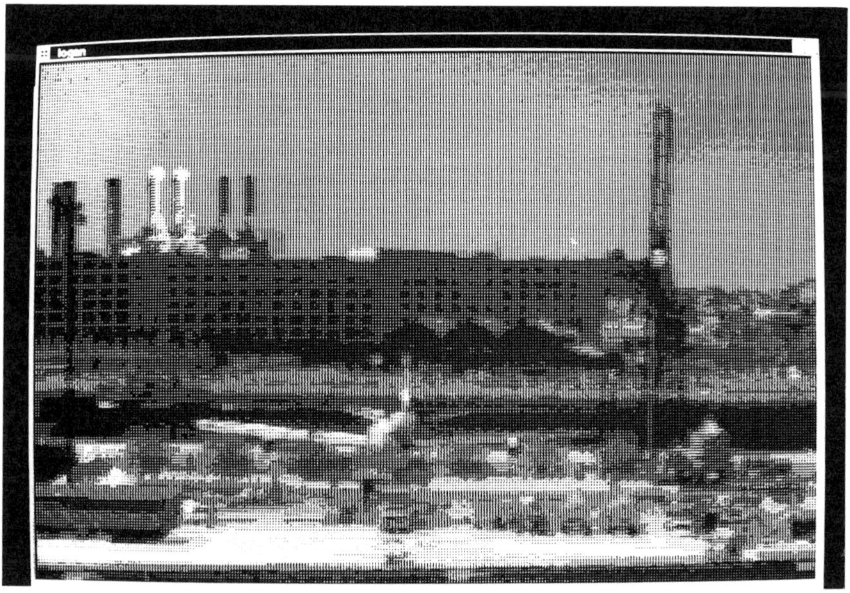

Figure 10.18 Iconographic composite picture of the visible and thermal images using the Color Icon. (See page 193 and Web site for color versions.)

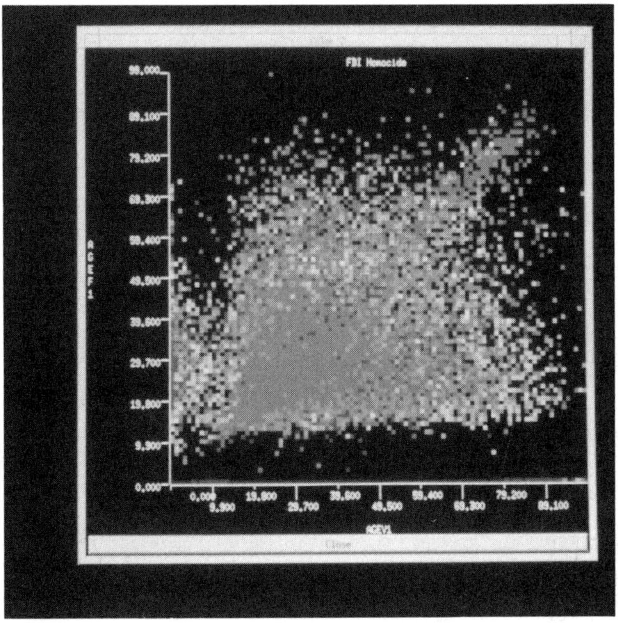

Figure 10.19 An integrated image of the FBI homicide database. See text for mapping details. (See Web site for color version.)

10.6.2 FBI homicide database

We have used the color icon to visualize an FBI database of homicide cases. This database stores information about homicide cases, such as the ages of the victim and the offender, the relationship (if any) between them, and the weapon used. Figure 10.19 is a scatterplot of this data. The age of the offender is mapped along the y axis. The age of the victim is mapped along the x axis.

We have used the following mappings. The color model used was the GLHS with weights $(0.5, 0.0, 0.5)$; this is the HLS double hexcone model, [LH93]. Limbs were mapped as follows:

Limb 1 L = constant, H = Sex of the victim, S = constant
Limb 2 L = constant, H = Sex of the offender, S = constant

with the following color key:

Top / Bottom	
Green / Green	Male Victim / Male Offender
Green / Blue	Male Victim / Female Offender
Blue / Blue	Female Victim / Female Offender
Blue / Green	Female Victim / Male Offender

Note the pattern of males killing females as the males get older. Examining the relationship field of the data shows that these are mostly husbands killing their wives.

10.6.3 Satellite images

We have also visualized satellite images of the great lakes area. The four raw satellite images are shown in black-and-white in Figure 10.20. Two integrated images, created using the color icon to merge the four raw images are shown in Figure 10.21. The top image was created with the GLHS color model using the weights (0.5, 0.0, 0.5) and the following mappings:

Limb 1	L = lake4, H = lake1, S = lake2
Limb 2	L = lake2, H = lake3, S = lake4

The bottom image of Figure 10.21 was created with the same GLHS color model and using the following mappings:

Limb 1	L = lake1, H = lake2, S = lake3
Limb 2	L = lake4, H = lake3, S = lake2

Figure 10.22 shows the use of the color icon to merge two three-parameter images. The original three-parameter images are shown at the top. A color icon image combining these two images is shown at the bottom. The resulting image has only four distinct parameters instead of a possible six because two of the parameters are redundant.

The first component image was created with the mapping Limb 1 $\mapsto$ L = lake1, H = lake2, S = lake3; the second one was create with the mapping Limb 1 $\mapsto$ L = lake4, H = lake3, S = lake2. The merged image was created with the mapping

Limb 1	L = lake1, H = lake2, S = lake3
Limb 2	L = lake4, H = lake3, S = lake2

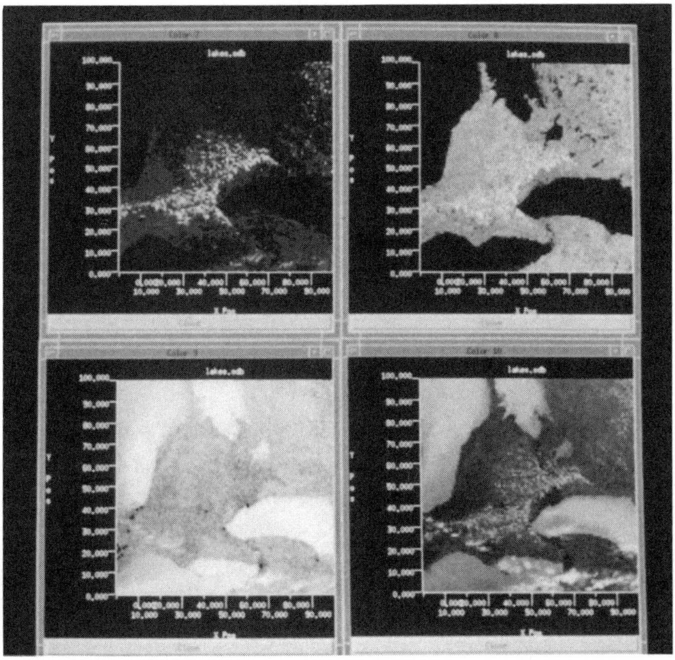

Figure 10.20 Four raw satellite images of the great lakes. (See also a version on the Web site.)

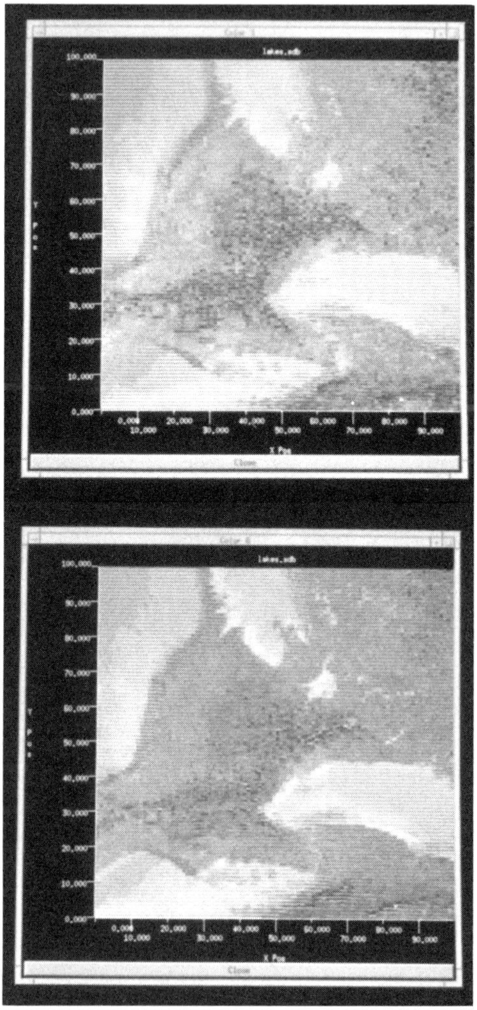

Figure 10.21 Two integrated images, created using the color icon to merge the four raw images using two different mappings, see text. (See Web site for color version.)

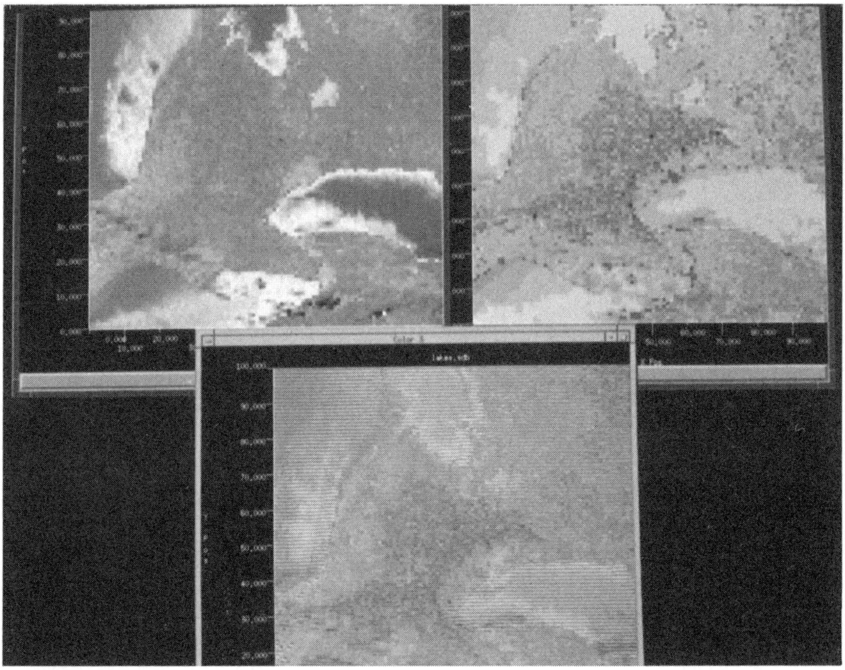

Figure 10.22 Merging two three-parameter images. The original three-parameter images are shown at the top. (See Web site for color version.)

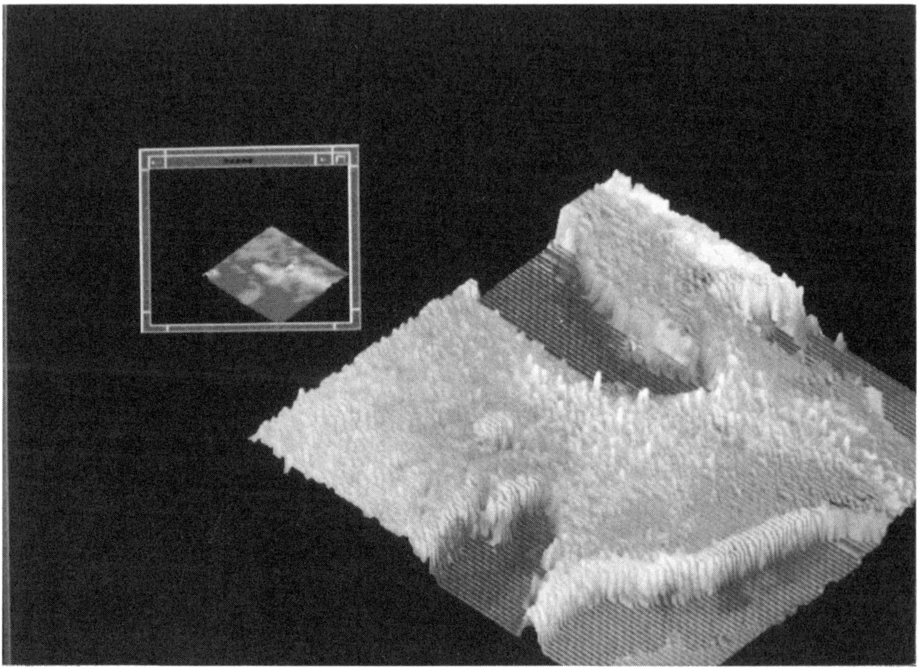

Figure 10.23 An example of a four-parameter color icon image with one of the parameters also mapped to height. (See page 193 and Web site for color versions.)

10.6.4 Color icon surface

Figure 10.23 shows an example of a four-parameter color icon image with one of the parameters also mapped to height.

10.6.5 Demonstrating the effect of controls with the parallel implementation

In a presentation to the Institute for Defense Analysis we were attempting to describe to individuals without strong backgrounds in computer graphics the effects of the different controls and color models that modify the results of the color icon. The parallel version of the color icon provided a mechanism by which they could get a feel for the effects rather than having to learn them only conceptually. The parallel version was of particular use when describing the GLHS color model, which these individuals had no experience with [LH93, LH88]. The color space description allowed them to attain a general idea of how the fields work. Being able to actually change the amount of lightness in the image and watch the image brighten and darken provided them with a truly useful understanding of the process. Hue and saturation worked especially well because they are much more difficult to conceptualize without experience.

This experience was very useful in that we were able to get feedback concerning the value of manipulating controls in real-time. One of our goals has been to provide a system where the individuals can make more intelligent decisions of the parameters or controls that need to be changed to affect the display in particular ways. This capability should help users generate more useful displays in less time.

10.7 SUMMARY AND NOTES

We have introduced a color-based icon, which employs both color and texture perception. Several options of features, attributes, and mappings of parameters to features and attributes have been presented.

We then described a new design of the color icon, a parallel implementation, and a few application examples. We have shown that a pseudo real-time implementation provides necessary and beneficial characteristics to the application, which makes it more comprehensible and therefore more useful.

The color icon technique could be easily extended also to a three-dimensional color voxel. This would allow the integration of multiple parameter data objects in three-dimensional scenes. Three-dimensional color icon surfaces and voxels could provide a method for conveying additional information in a virtual environment. Color icon voxels could be used to render multiple parameter volumes. Further three-dimensional extensions can utilize stereo disparity to convey parameter values.

Work needs to be done to develop a true real-time implementation. The primary task here would be to improve the display time so that it more closely matches calculation time. In addition, additional hardware types need to be explored, particularly MIMD systems, in order to assess whether any speedup over SIMD execution can be gained.

Since the color icon uses the same interpolation method as Gouraud shading, one possible alternative is a very fast implementation using existing three-dimensional graphics hardware. For comparison, the Terasys implementation is drawing a 64×64 icon image. This would correspond to a 64×64 point quadrilateral mesh. This type of a mesh can be implemented very simply and efficiently in many existing three-dimensional rendering libraries such as PEX and GL. Implementations of the color icon on accelerated graphics boards may come very close to the performance of the parallel implementation. This method would be very attractive to most people because it would be using hardware that many already have. In addition, it would make the use of color icons in existing visualization applications very simple.

Last, the entire approach needs to be examined more rigorously with test subjects to determine its strengths and weaknesses. We outline such evaluations now.

10.7.1 Evaluation studies

As we have described and demonstrated above, there are many different possibilities to generate integrated display. From a technological point of view most of them are feasible. However, it is not possible to determine their efficacy without evaluation studies with human observers.

The first step towards evaluation is selecting and adjusting the various imaging variables so as to explore the different combinations that are possible. In such a study, the participants try different viewing variables of the integrated imaging system. Since the number of combinations can be very large, the goal is to try to identify those particular selections and values that appear to be most promising. Those choices are then used for subsequent, more elaborate studies. Such studies should employ specific tasks such as detection, classification, localization, and quantification of objects; larger sample sizes; larger numbers of participants; and more rigorous analysis methods.

When designing such studies, both real images and synthesized data should be considered. Real data present real-world problems but may lack information about the "truth." The contents of synthesized data is always known, and can be adjusted to test situations for which real data is not available, but results from studies using synthesized data may not extend to the real world.

11

COLOR ON THE WORLD-WIDE WEB

The World-Wide Web has become a major source of information. Not only does it provide a wealth of information for you to consume, but it also gives you the opportunity to be an information producer, a publisher.

This, in turn, gives you the opportunity to violate every known recommendation for effective of color usage! Indeed, a large majority of information producers on the Web either have not bothered to study such recommendations, or have decided to deliberately ignore them.

Color on the Web has experienced the same neglect as in other visual computing applications, such as computer graphics, imaging, and visualization. In all these applications, color has been ignored or neglected for a long period of time before interest has mounted.

A case in point is the default color most Web browsers assign to hyperlinks: blue. As we have seen earlier, our visual system has the lowest sensitivity to blue, making it the worst color for small samples, such as text and line graphics.

There are many other examples of sites on the Web that burden the viewer with text in colors that are bearly readable, with insufficient contrast between foreground and background, with annoying color and texture combinations, and more.

This chapter briefly explores color issues on the World-Wide Web. We summarize the main color capabilities that are available to us on the Web today; the summary is as short as the capabilities to date are.

We anticipate that, as with other visual computing applications, eventually interest will arise in using and supporting color in a more complete an effective way. We thus proceed to outline what we consider to be the needs still ahead to obtain the improved color support needed.

11.1 COLOR CAPABILITIES ON THE WEB

Unfortunately, when you are facing the question what color schemes to use on your World-Wide Web site, you do not have too many options. This will make your decision relatively simple, but might limit your ability to deliver your information in what may be the optimal way.

Color support by Web browsers is extremely limited. All graphical browsers support CompuServe's GIF image format. Additionally, most modern browsers will support JPEG formats. That's about it.

11.1.1 The CompuServe GIF Format

CompuServe's Graphics Interchange Format (GIF) is the most popular image format on the Web. The format can be used to store multiple bitmap images in a single file for exchange among different systems and platforms, though it is mostly used on the Web in a single image per file mode.

GIF image files can carry up to eight bits of color, that is, up to 256 colors. While sufficient for many application, this is a severe limit on the color contents an image on the Web can have.

Other technical details of the format are beyond the scope of this book. For more details, see [Com87, Com90, Mv94].

11.1.2 JPEG: The Joint Photographic Experts Group File Interchange Format

JPEG is used to refer to the Joint Photographic Experts Group standards organization, the file compression method developed by that organization, and the file format used in this method.

JPEG compression is capable of compressing continuous-tone images with six to 24 bits per pixel. JPEG does it relatively fast and efficient.

This stands in contrast to other formats, including GIF, which do not do as good a job compressing continuous-tone images.

JPEG defines neither a single standard file format nor a single algorithm. It is a collection of compression methods that can be used to meet the needs at hand. Several file formats have been developed to accommodate the needs of the JPEG storage specifications.

While this presents a powerful set of tools, it also puts heavy demands on the receiving end, to be able to decode all the possible file formats, using all the possible compression algorithms that might have been used to compress and store the coming image.

One of the strong points of JPEG color compression is that, while using a lossy compression scheme, it preserves loseless luminance information, channeling all the lossy encoding to the chrominance components. As we have seen in Chapter 1, this is the most appropriate compression scheme, as viewers are much more likely to detect losses in luminance than in chrominance.

However, JPEG does not provide good results for all types of images. The approach was designed to compress images of real-world subjects. It does not compress well black-and-white images, line art, vector graphics, ray-traced images, or animations. Additionally, large areas of a single color end up with artifacts the degrade them significantly.

Other shortcomings of JPEG include the need for a 24-bit display to obtain satisfactory results; its poor performance when implemented in software; its complex implementation; and its lack of support by many file formats.

All in all, while presenting desirable features theoretically, they are not practical for a large majority of Web users.

For more details, see [ISOa, ISOb, Mv94].

11.2 WHAT'S AHEAD?

Web color support needs to catch up with general graphics support. As more and more computers possess better graphics capabilities, Web browsers should offer better support for higher color resolutions.

Future browsers will need to handle typical graphics color resolutions of four channels (Red, Green, Blue, and Alpha), at 8 bits each, at least.

Color specifications should become more "user-friendly," and color management and manipulation capabilities should be incorporated into browsers. Such software would allow users to adjust color parameters of images they are viewing on the Web. Alternatively, one could develop, e.g., applets, which users could download to view and manage better color images (similar to the way PDF viewers are currently available to view Adobe Acrobat files).

Browsers should become more "intelligent," and should be able to harness built-in hardware (such as graphics hardware) on the client's machine. This would allow better implementation of JPEG schemes.

In addition, new file formats should be developed (whether within the JPEG framework or otherwise) that will provide better support for better color resolutions (and other imaging parameters). One approach could define a special extension of the file header to include built-in applets, which would be included in the image file to provide the appropriate de-compression algorithms, as well as image processing and color management tools. Alternatively, such filters and image processing capabilities can be incorporated in the browser.

These possibilities provide a rich spectrum of options for research and development, and should attract researcher and developer, both in academia and industry.

Appendix A. Color Illustrations

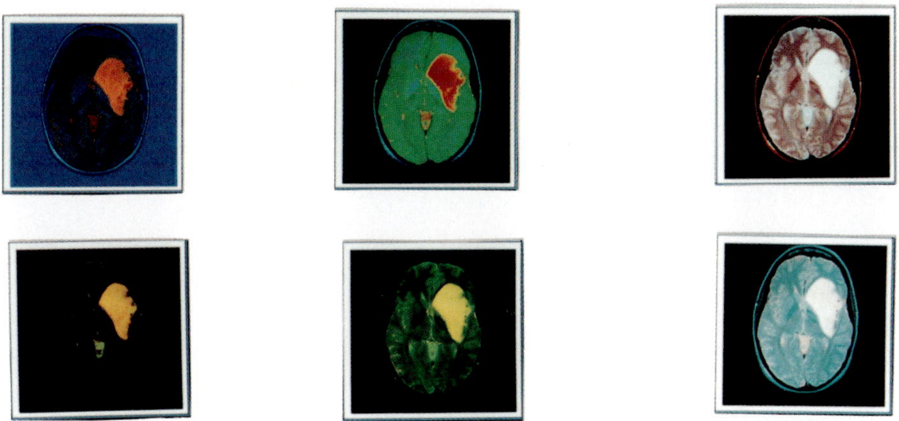

(a) (b) (c)

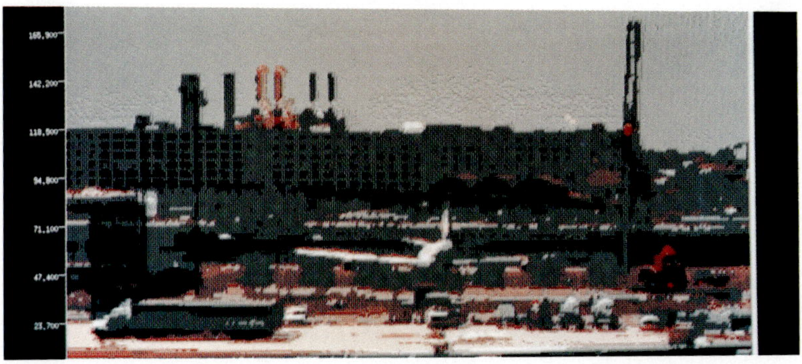

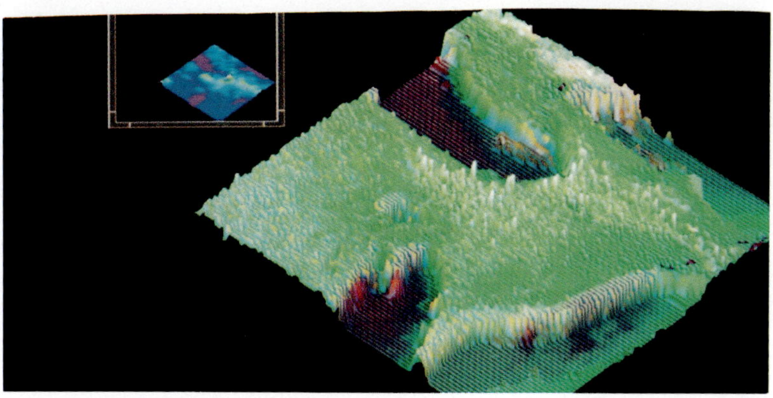

REFERENCES

[AGC90] I. Abramov, J. Gordon, and H. Chan. Using hue scaling to specify color appearance and to derive color differences. In M. H. Brill, editor, *Proceedings of the SPIE, Volume 1250, Perceiving, Measuring, and Using Color*, pages 40–53. 1990.

[Ago79] G. A. Agoston. *Color Theory and Its Application in Art and Design.* Springer-Verlag, Berlin and Heidelberg, West Germany, 1979.

[AHU76] A. V. Aho, J. E. Hopcroft, and J. D. Ullman. *The Design and Analysis of Computer Algorithms.* Addison-Wesley, Reading, MA, 1976.

[Alb75] J. Albers. *Interaction of Color.* Yale University Press, New Haven, CT, 1975.

[Are91] L. Arend. Apparent contrast and surface color in complex scenes. In M. H. Brill B. E. Rogowitz and J. P. Allebach, editors, *Proceedings of the SPIE, Volume 1453, Human Vision, Visual Processing and Digital Display II*, pages 412–421. 1991.

[BBK82] T. Berk, L. Brownston, and A. Kaufman. A new color-naming system for graphics languages. *IEEE Computer Graphics and Applications*, 2(3):37–44, May 1982.

[Boy89] R. M. Boynton. Eleven colors that are almost never confused. In B. E. Rogowitz, editor, *Proceedings of the SPIE, Volume 1077, Human Vision, Visual Processing and Digital Display*, pages 322–332. 1989.

[BPR82] J. Beck, K. Prazdny, and A. Rosenfeld. A theory of textural segmentation. In J. Beck, B. Hope, and A. Rosenfeld, editors, *Human and Machine Vision*, pages 1–38. Academic Press, New York, 1982.

[CIE78] CIE: Commission Internationale de l'Eclairage. CIE recommendations on uniform color spaces—color difference equations, psychometric color terms. *CIE Publication*, (15, (E-1.3.1) 1971/(TC-1.3) 1978, Supplement No. 2):9–12, 1978.

[Com87] CompuServe Incorporated. GIF Graphics Interchange Format: A standard defining a mechanism for the storage and transmission of bitmap-based graphics. Columbus, OH, 1987.

[Com90] CompuServe Incorporated. Graphics Interchange Format: Version 89a. Columbus, OH, 1990.

[Cow83] W. B. Cowan. An inexpensive scheme for calibration of a color monitor in terms of CIE standard co-ordinates. *Computer Graphics*, 17(3):315–321, July 1983. Proceddings ACM SIGGRAPH.

[DA69] D. D. Dorfman and E. Alf, Jr. Maximum likelihood estimation of parameters of signal-detection theory and determination of confidence intervals–rating method data. *Journal of Mathematical Psychology*, 6:487–496, 1969.

[EGL95a] R. Erbacher, D. Gonthier, and H. Levkowitz. The color icon: A new design and a parallel implementation. In G. Grinstein and R. Erbacher, editors, *Proceedings of the SPIE 95 Conference on Visual Data Exploration and Analysis II*, pages 302–312, San Jose, CA, February 5–10 1995. SPIE.

[EGL+95b] R. F. Erbacher, G. G. Grinstein, J. P. Lee, H. Levkowitz, L. Masterman, R. Pickett, and S. Smith. Exploratory visualization research at the university of massachusetts at lowell. *Computer and Graphics*, 19(1):131–139, 1995.

[Enn88] J. T. Enns. Three-dimensional features that pop out in visual search. In *Proceedings of the First International Conference on Visual Search*, Durham, UK, 1988.

[Enn90] J. T. Enns. Three-dimensional features that pop out in visual search. In D. Brogan, editor, *Visual Search*. Taylor and Francis, London, 1990.

[Enn91] J. T. Enns. The nature of selectivity in early human vision. In B. Burns, editor, *Percepts, Concepts, and Categories: The Representation and Processing of Information*. Elsevier Science Publications, Amsterdam, 1991. To appear.

[Far83] E. J. Farrell. Color display and interactive interpretation of three-dimensional data. *IBM Journal of Research and Development*, 27(4):356–366, July 1983.

[Fel89] U. Feldman. Tulip, a modified Munsell color space. In *Intelligent Robots and Computer Vision*, Philadelphia, PA, November 1989. SPIE.

[Fis83] K. P. Fishkin. *Applying color science to computer graphics.* PhD thesis, Computer Science Division, University of California, Berkeley, 1983.

[FvD82] J. D. Foley and A. van Dam. *Fundamentals of Interactive Computer Graphics.* Addison-Wesley, Reading, MA, 1982.

[FvFH90] J. D. Foley, A. van Dam, S. K. Feiner, and J. F. Hughes. *Computer Graphics Principles and Practice.* Addison-Wesley, Reading, MA, second edition, 1990.

[Gib50] J. J. Gibson. *The Perception of the Visual World.* Houghton-Mifflin, Boston, MA, 1950.

[GS90] G. Grinstein and S. Smith. The perceptualization of scientific data. In E. J. Farrell, editor, *Extracting Meaning from Complex Data: Processing, Display, Interaction*, pages 190–199, Bellingham, WA, 1990. SPIE—The International Society for Optical Engineering.

[Gut89] S. L. Guth. Unified model for human color perception and visual adaptation. In B. E. Rogowitz, editor, *Proceedings of the SPIE, Volume 1077, Human Vision, Visual Processing and Digital Display*, pages 370–390. 1989.

[Hes84] S. Hesselgren. Why color order systems? *COLOR research and application*, 9(4):220–228, Winter 1984.

[HJ57] L. M. Hurvich and D. Jameson. An opponent-process theory of color vision. *Psychological Review*, 64:384–390, 1957.

[HS81] A. Hard and L. Sivik. NCS-natural color system: A Swedish standard for color notation. *COLOR research and application*, 6(3):129–138, Fall 1981.

[Hub88] D. Hubel. *Eye, Brain, and Vision.* W. H. Freeman and Co., New York, NY, 1988.

[Hun77] R. W. G. Hunt. The specification of colour appearance I. Concepts and terms. *Color research and application*, 2:55–68, 1977.

[Hun87] R. W. G. Hunt. *Measuring Colour.* Ellis Horwood, Ltd., 1987.

[ISOa] ISO: International Standards Organization. Digital compression and coding of continuous-tone still images, part 1: Requirements and guidelines. Document number ISO/IEC IS 10918-1.

[ISOb] ISO: International Standards Organization. Digital compression and coding of continuous-tone still images, part 2: Compliance testing. Document number ISO/IEC CD 10918-2.

[JB89] N. Jacobson and W. Bender. Strategies for selecting a fixed palette of colors. In *Human Vision, Visual Processing and Digital Displays*, pages 333–341, CA, January 1989. SPIE.

[JG78] G. H. Joblove and D. Greenberg. Color spaces for computer graphics. *Computer Graphics*, 12(3):20–25, August 1978.

[JW75] D. B. Judd and G. Wyszecki. *Color in Business, Science, and Industry.* J. Wiley and Sons, New York, NY, 1975. 3rd. ed.

[KK82] P. R. Keller and M. Keller. *Visual Cues.* IEEE Computer Society Press, Los Alamitos, California, 1982.

[Lev88] H. Levkowitz. *Color in Computer Graphic Representation of Two-Dimensional Parameter Distributions.* PhD thesis, Department of Computer and Information Science, The University of Pennsylvania, Philadelphia, PA, August 1988. (Technical Report MS-CIS-88-100, Department of Computer and Information Science and MIPG139, Medical Image Processing Group, Department of Radiology, University of Pennsylvania.).

[Lev91] H. Levkowitz. Color icons: Merging color and texture perception for integrated visualization of multiple parameters. In G. M. Nielson and L. J. Rosenblum, editors, *Visualization '91*, pages 164–170, San Diego, CA, October 22–25 1991. IEEE Computer Society, IEEE Computer Society Press.

[Lev96] H. Levkowitz. Perceptual steps along color scales. *International Journal of Imaging Systems and Technology*, 7:97–101, 1996.

[LH86] H. Levkowitz and G.T. Herman. Color in multidimensional multiparameter medical imaging. *COLOR research and application*, 11(Supplement):S15–S20, 1986.

[LH87a] H. Levkowitz and G.T. Herman. GIHS: A generalized color model and its use for the representation of multiparameter medical images. In M.A. Viergever and A.E. Todd-Pokropek, editors, *Mathematics and Computer Science in Medical Imaging*, pages 389–399. Springer-Verlag, Berlin, 1987.

[LH87b] H. Levkowitz and G.T. Herman. Towards an optimal color scale. In *Computer Graphics '87*, pages 92–98, Philadelphia, PA, March 22–26 1987. National Computer Graphics Association.

[LH88] H. Levkowitz and G.T. Herman. Towards a uniform lightness, hue, and saturation color model. In *Electronic Imaging Devices and Systems '88: Image Processing, Analysis, Measurement, and Quality*, pages 215-222. SPSE—The Society for Imaging Science and Technology, January 10–15 1988.

[LH92] H. Levkowitz and G. T. Herman. Color scales for image data. *IEEE Computer Graphics and Applications*, 12(1):72–80, January 1992.

[LH93] H. Levkowitz and G. T. Herman. GLHS: A generalized lightness, hue, and saturation color model. *CVGIP: Graphical Models and Image Processing*, 55(4):271–285, July 1993.

[LP90] H. Levkowitz and R. M. Pickett. Iconographic integrated displays of multiparameter spatial distributions. In B. E. Rogowitz and J. P. Allebach, editors, *SPIE '90. Human Vision and Electronic Imaging: Models, Methods, and Applications*, pages 345–355, Santa Clara, CA, February 12–14 1990.

[LR86] M. R. Luo and B. Rigg. Uniform colour space based on the CMC(l : c) colour-difference formula. *Journal of the Society of Dyers and Colourists*, 102:164–172, May/June 1986.

[LR87a] M. R. Luo and B. Rigg. BFD(l : c) colour-difference formula: Part 1—development of the formula. *Journal of the Society of Dyers and Colourists*, 103:86–94, February 1987.

[LR87b] M. R. Luo and B. Rigg. BFD(l : c) colour-difference formula: Part 2—performance of the formula. *Journal of the Society of Dyers and Colourists*, 103:126–132, March 1987.

[LR87c] M. R. Luo and B. Rigg. A colour-difference formula for surface colours under illuminant A. *Journal of the Society of Dyers and Colourists*, 103:161–168, April 1987.

[LSPD90] H. Levkowitz, S. E. Seltzer, R. M. Pickett, and C. J. D'Orsi. Color integrated displays of multimodality tomographic images. *Radiology*, 177 (P):248, November 25–30 1990. Poster presented at the 76th Annual Meeting of the Radiological Society of North America (RSNA '90).

[LX92] H. Levkowitz and L. L. Xu. Approximating the Munsell Book of Color with the Generalized Lightness, Hue, and Saturation color model. In *SPIE '92*, San Jose, CA, February 9–14 1992. To appear.

[Mac43] D. L. MacAdam. Specification of small chromaticity differences. *Journal of the Optical Society of America*, 33(2):18–26, 1943.

[Mar82] D. Marr. *Vision*. W. H. Freeman and Company, New York, NY, 1982.

[MG80] G. W. Meyer and D. P. Greenberg. Perceptual color spaces for computer grapohics. *Computer Graphics*, 14(3):254–261, July 1980. Proceedings, SIGGRAPH 1980.

[MG87] G. W. Meyer and D. P. Greenberg. Perceptual color spaces for computer graphics. In H. J. Durrett, editor, *Color and the Computer*, chapter 4, pages 83 – 100. Academic Press, Orlando, FL, 1987.

[MG88] G. W. Meyer and D. P. Greenberg. Color-defective vision and computer graphics displays. *IEEE Comptuer Graphics and Applications*, 8(5):28–40, September 1988.

[MN93] J. Marsh and M. Norder. Pim chip specification. Technical Report SRC-TR-93-088, Super Computing Research Center, Institute for Defense Analyses, Bowie, MD, 1993.

[Mun05] A. H. Munsell. *A Color Notation*. Munsell Color Company, Boston, MA, 1905.

[Mun69] A. H. Munsell. *A Grammar of Color*. Van Nostrand Reinhold Company, New York, NY, 1969.

[Mun76] Munsell Color Company. *The Munsell Book of Color*. Munsell Color Company, 2441 North Calvert Street, Baltimore, MD 21218, 1976. Under continuous update.

[Mur84a] G. M. Murch. The effective use of color: cognitive principles. *Tekniques*, 8(2):25–31, 1984.

[Mur84b] G. M. Murch. The effective use of color: perceptual principles. *Tekniques*, 8(1):4–9, 1984.

[Mur84c] G. M. Murch. Physiological principles for the effective use of color. *IEEE Computer Graphics and Applications*, 4(11):49–54, November 1984.

[Mv94] J. D. Murray and W. vanRyper. *Encyclopedia of Graphics File Formats*. O'Reilly and Associates, Inc., Sebastopol, CA, 1994.

[Nem80] A. Nemcsics. The colorid color system. *COLOR research and application*, 5(2):113–120, Summer 1980.

[New04] Sir Isaac Newton. *Opticks*. Sam Smith and Benjamin Walford, London, 1704. See also Dover Publications, Inc., New York, N.Y. 1952.

[NS86] K. Nakayama and G. H. Silverman. Serial and parallel processing of visual feature conjunctions. *Nature*, 320:264–265, 1986.

[Ost31] W. Ostwald. *Colour Science*. Windsor and Newton, London, 1931.

[Ost69] W. Ostwald. *The Color Primer*. Van Nostrand Reinhold Company, New York, NY, 1969.

[PG88] R.M. Pickett and G. Grinstein. Iconographic displays for visualizing multidimensional data. In *Proceedings of the 1988 IEEE Conference on Systems, Man and Cybernetics*, Beijing and Shenyang, People's Republic of China, 1988.

[Pic70] R. M. Pickett. Visual analyses of texture in the detection and recognition of objects. In B. S. Lipkin and A. Rosenfeld, editors, *Picture Processing and Psycho-Pictorics*. Academic Press, New York, 1970.

[Piz81a] S. M. Pizer. Intensity mappings: Linearization, image-based, user-controlled. *SPIE*, 271(Display Technology II):21–27, 1981.

[Piz81b] S. M. Pizer. Intensity mappings to linearize display devices. *Computer Graphics and Image Processing*, 17(3):262–268, 1981.

[PLS90] R. M. Pickett, H. Levkowitz, and S. Seltzer. Iconographic displays of multiparameter and multimodality images. In *Proceedings of the First Conference on Visualization in Biomedical Computing*, pages 58–65, Atlanta, GA, May 22–25 1990. IEEE Computer Society Press.

[PZJ82] S. M. Pizer, J. B. Zimmerman, and R. E. Johnston. Contrast transmission in medical image display. In *Proceedings of the 1st International Symposium on Medical Imaging and Interpretation ISMIII'82*, pages 2–9, October 1982.

[RO85] P. K. Robertson and J. F. O'Callaghan. The application of scene synthesis techniques to the display of multidimensional image data. *ACM Transactions on Graphics*, 1(4):274–288, October 1985.

[RO86] P. K. Robertson and J. F. O'Callaghan. The generation of color sequences for univariate and bivariate mapping. *IEEE Computer Graphics and Applications*, 6(2):24–32, February 1986.

[Rob84] A. R. Robertson. Colour order systems: An introductory review. *COLOR research and application*, 9(4):234–240, Winter 1984.

[Rob85] P. K. Robertson. *Colour Image Display: Computational Framework Based on a Uniform Colour Space*. PhD thesis, Australian National University, April 1985. CSIRONET Tech. Rep. No. 27.

[Rob88] P. K. Robertson. Visualising color gamuts: A user interface for the effective use of perceptual color spaces in data displays. *IEEE Computer Graphics and Applications*, 8(5):50–64, September 1988.

[RP89] B. E. Rogowitz and J. Park. A descriptive model of display flicker. IBM Research Report RC 15901 (#68186), 1989.

[Run73] O. P. Runge. *Die Farbenkugel*. Bern Benteli, England, 1973. Originally published 1810, Germany.

[SB90] R. Sekuler and R. Blake. *Perception*. McGraw-Hill Publishing Company, New York, NY, second edition, 1990.

[SBCG84] M. W. Schwartz, J. C. Beatty, Wm. B. Cowan, and J. F. Gentleman. Towards an effective user interface for interactive colour manipulation. *Graphics Interface*, pages 187–196, May 1984.

[SCB87] M. W. Schwartz, Wm. B. Cowan, and J. C. Beatty. An experimental comparison of RGB, YIQ, LAB, HSV, and Opponent Color Models. *ACM Transactions on Graphics*, 6(2):123–158, April 1987.

[SCB88] M. C. Stone, Wm. B. Cowan, and J. C. Beatty. Color gamut mapping and the printing of digital color images. *ACM Transaction on Graphics*, 7(3):249–292, October 1988.

[Sch56] O. H. Schade. Optical and photoelecctric analog of the eye. *Journal of the Optical Society of America*, 46:721–739, 1956.

[Sch87] R. A. Schuchard. *Evaluation of pseudocolor and achromatic CRT display scales in medical imaging*. PhD thesis, University of Chicago, 1987.

[Sch90a] H. R. Schiffman. *Sensation and Perception*. John Wiley and Sons, Somerset, NJ, 3rd edition, 1990.

[Sch90b] R. A. Schuchard. Review of colorimetric methods for developing and evaluating uniform CRT display scales. *Optical Engineering*, 29(4):378–384, April 1990.

[SGB91] S. Smith, G. Grinstein, and R. D. Bergeron. Interactive data exploration with a supercomputer. In *Proceedings of Visualization '91*, San Diego, CA, 1991.

[Smi78a] A. R. Smith. Color gamut transform pairs. *Computer Graphics*, 12(3):12–19, August 1978.

[Smi78b] A. R. Smith. Realizable colors. Technical Memo 8, Computer Graphics Lab, New York Institute of Technology, August 1978.

[SP82] J. A. Swets and R. M. Pickett. *Evaluation of Diagnostic Systems*. Academic Press, New York, N.Y., 1982.

[SS92] S. Smith and L. A. Scarff. Combining visual and IR images for sensor fusion: two approaches. In *Proceedings of the SPIE/IS&T Conference on Electronic Imaging*, San Jose, CA, 1992.

[Str35] J. R. Stroop. Studies of interference in serial adverbal reactions. *Journal of Experimental Psychology*, 18:643–662, 1935.

[Swe93] H. D. Sweely. Terasys demonstration hardware manual. Technical Report SRC No. 101. 93, Super Computing Research Center Institute for Defense Analyses, Bowie, MD, 1993.

[Taj83a] J. Tajima. Optimal color display using uniform color scale. *NEC Research and Development*, (70):58–63, July 1983.

[Taj83b] J. Tajima. Uniform color scale applications to computer graphics. *Computer Vision, Graphics, and Image Processing*, 21(3):305–325, March 1983.

[TG80] A. E. Triesman and G. A. Gelade. A feature integration theory of attention. *Cognitive Psychology*, 12:97–136, 1980.

[TG88] A. Treisman and S. Gormican. Feature analysis in early vision: Evidence from search asymmetries. *Psychological Review*, 95:14–48, 1988.

[TM86] J. M. Taylor and G. M. Murch. The effective use of color in visual displays: Text and graphics applications. *COLOR research and application*, 11(Supplement):S3–S10, 1986.

[TMM89] J. M. Taylor, G. M. Murch, and P. A. McManus. TekHVC^{tm}: A uniform perceptual color system for display users. *Proceedings of the SID*, 30(1):15–21, 1989.

[Tod83] A. E. Todd-Popkropek. The intercomparison of a black and white and a color display: An example of the use of receiver operating characteristic curves. *IEEE Transactions on Medical Imaging*, MI-2(1):19–23, March 1983.

[TS90] L. G. Thorell and W. J. Smith. *Using Computer Color Effectively: An Illustrated Reference.* Prentice Hall, New Jersey, 1990.

[von66] H. L. F. von Helmholtz. *Handbuch der Physiologischen Optik.* Voss, Hamburg, Germany, 1866. 3rd ed. translated by J. P. C. Southball, Optical Society of America, 1924-25, Dover Publication, New York, NY, 1962.

[von70] J. W. von Goethe. *Theory of Colours.* MIT Press, Cambridge, MA, 1970. English trans. C. L. Eastlake, London, 1840. First ed. 1810.

[W. 80] W. Morris, (ed.). The american heritage dictionary of the english language. Houghton Mifflin Company, Boston, 1980.

[Wan95] B. A. Wandell. *Foundations of Vision.* Sinauer Associates, Inc., Sunderland, MA, 1995.

[War88] C. Ware. Color sequences for univariate maps: Theory, experiments, and principles. *IEEE Computer Graphics and Applications*, 8(5):41–49, September 1988.

[WBR87] J. Walraven, T. L. Benzschawel, and B. E. Rogowitz. Color-constancy interpretation of chromatic induction. Die Farbe, 34:269–273, 1987.

[WC90] C. Ware and Wm. B. Cowan. The RGYB color geometry. *ACM Transactions on Graphics*, 9(2):226–232, April 1990.

[Wri84] W. D. Wright. The basic concepts and attributes of colour order systems. *COLOR research and application*, 9(4):229–233, Winter 1984.

[WS67] G. Wyszecki and W. S. Stiles. *Color Science.* J. Wiley and Sons, NY, 1967.

[Xu96] L. Xu. *Uniform Color Appearance Model for Color Reproduction in Graphic Art.* PhD thesis, Department of Computer Science, University of Massachusetts Lowell, December 1996.

ADDITIONAL BIBILIOGRAPHY

[1] E. H. Adelson and J. R. Bergen. Spatiotemporal energy models for the perception of motion. *Journal of the Optical Society of America, A*, 2:284–299, February 1985.

[2] J. Aloimonos and D. Shulman. *Integration of Visual Modules: An extension of the Marr paradigm.* Academic Press, Boston, MA, 1989. ISBN 0-12-053020-1.

[3] M. A. Andreottola. Color hard-copy devices. In H. J. Durrett, editor, *Color and the Computer*, chapter 12, pages 221–240. Academic Press, Orlando, FL, 1987.

[4] T. Benzschawel and S. L. Guth. Atdn: Toward a uniform color space. *COLOR research and application*, 9(4):133–141, Winter 1984.

[5] F. J. Bouma. *Physical Aspects of Colour.* N. V. Philips Gloeilampenfabriken, Eindhoven, Netherlands, 1974.

[6] P. T. Breen, Miller-Jacobs P. E., and H. H. Miller-Jacobs. Color displays applied to command, control, and communication (c^3) systems. In H. J. Durrett, editor, *Color and the Computer*, chapter 9, pages 171–188. Academic Press, Orlando, FL, 1987.

[7] G. Buchsbaum and S. D. Bedrosian. Number of simultaneous colors versus gray levels: a quantitative relationship. *Proceedings of IEEE*, 72(10):1419–1421, October 1984.

[8] G. Buchsbaum and A. Gottschalk. Trichromacy, opponent colours coding and optimum colour information in the retina. *Proceedings of the Royal Society of London*, B220:89–113, 1983.

[9] J. J. Chang and J. D. Carroll. Three are not enough: An indscal analysis suggesting that color has seven (± 1) dimensions. *COLOR research and application*, 5(4):193–206, Winter 1980.

[10] M. E. Chevreul. *The Principles of Harmony and Contrast of Colors and Their Applications in the Arts.* Van Nostrand Reinhold, New York, N.Y., 1957. Reprinted. Original version published 1839.

[11] J. F. Cornhill, W. A. Barrett, E. E. Herderick, R. W. Mahley, and D. L. Fry. Topographic study of sudanophilic lesions in cholesterol-fed minipigs by image analysis. *Arteriosclerosis*, 5(5):415–426, September/October 1985.

[12] W. B. Cowan, editor. *Proceedings of the 1986 AIC Interim Meeting on Color in Computer Generated Displays*, Toronto, Ontario, Canada, June 19–20 1986. Color Research and Application. Volume 11 Supplement.

[13] W. B. Cowan. Colour psychophysics and display technology: Avoiding the wrong answers and finding the right questions. In G. H. Hugnes, P. E. Mantey, B. E. Rogowitz, editors, *Proceedings of the SPIE, Volume 901, Image Processing, Analysis, Measurement and Quality*, 901, 186–193, 1988.

[14] C. M. M. de Weert. Superimposition of colour information. *COLOR research and application*, 11(Supplement):S21–S26, 1986.

[15] H. J. Durrett and D. T. Stimmel. Color and the instructional use of the computer. In H. J. Durrett, editor, *Color and the Computer*, chapter 13, pages 241–254. Academic Press, Orlando, FL, 1987.

[16] H. J. Durrett (ed.). *Color and the Computer.* Academic Press, Orlando, FL, 1987.

[17] M. Faintich. Science or art? *Computer Graphics World*, pages 81–86, April 1985.

[18] E. J. Farrell, R. Zappulla, and W. C. Yang. Color 3-D imaging of normal and pathologic intracranial structures. *IEEE Computer Graphics and Applications*, 4(9):5–17, September 1984.

[19] F. Gerristen. *Theory and Practice of Color.* Van Nostrand Reinhold Company, New York, NY, 1975.

[20] R. Gershon. Aspects of perception and computation in color vision. *Computer Vision, Graphics, and Image Processing*, 32(2):244–277, November 1985.

[21] R. Gershon. *The use of color in computational vision.* PhD thesis, University of Toronto, Toronto, Ontario, Canada, June 1987. Technical Report RBCV-TR-87-15.

[22] D. J. Gilderdale, J. M. Pennock, and E. M. Bydder. Clinical applications of colour display in NMR imaging. In *Book of Abstracts, 4th Annual Meeting*, pages 1234–1235, London, UK, 19–23 August 1985. Society of Magnetic Resonance Imaging in Medicine.

[23] W. E. Glenn and D. G. Glenn. The design of systems that display moving images based on spatio-temporal vision data In G. H. Hugnes, P. E. Mantey, B. E. Rogowitz, editors, *Proceedings of the SPIE, Volume 901, Image Processing, Analysis, Measurement and Quality*, 901, 230–240, 1988.

[24] P. Gouras. Visual system iv: Color vision. In *Principles of Neural Science*. Elsvier/North-Holland, New York/Amsterdam, 1981.

[25] F. Grum and C. J. Bartleson (eds.). *Optical Radiation Measurements*, volume 2—Color Measurement. Academic Press, New York, N.Y., 1980.

[26] G. T. Herman and H. Levkowitz. Contributions to the NATO advanced study institute. Technical Report MIPG116, Medical Image Processing Group, Department of Radiology, University of Pennsylvania, Philadelphia, PA 19104, Month 1987.

[27] G.T. Herman and H. Levkowitz. Color scales for medical image data. In *Electronic Imaging 88*, pages 1068–1073, Boston, MA, October 3–6 1988. Institute for Graphic Communication.

[28] R. W. G. Hunt. *The Reproduction of Color*, volume 3rd ed. J. Wiley and Sons, New York, N.Y., 1975.

[29] L. M. Hurvich and D. Jameson. An opponent-process theory of color vision. *Psychological Review*, 64:384–390, 1957.

[30] N. E. Jacobs. The executive decision: selecting a business graphics system. In H. J. Durrett, editor, *Color and the Computer*, chapter 16, pages 285–294. Academic Press, Orlando, FL, 1987.

[31] C. C. Jaffe. Color displays: widening imaging's dynamic range. *Diagnostic Imaging*, 7(12):52–58, December 1985.

[32] D. H. Kelly. Spatial and temporal aspects of red/green opponency In G. H. Hugnes, P. E. Mantey, B. E. Rogowitz, editors, *Proceedings of the SPIE, Volume 901, Image Processing, Analysis, Measurement and Quality*, 901, 258–267, 1988.

[33] H. Levkowitz (Chair). Panel: Color vs. black-and-white in visualization. In *Visualization '91*, San Diego, CA, October 22–25 1991. IEEE Computer Society.

[34] H. Levkowitz and G. G. Grinstein. Experimental approaches to color. In *Electronic Imaging 90*, Boston, MA, October 29–November 1 1990.

[35] H. Levkowitz and G.T. Herman. Color in multidimensional multiparameter medical imaging. *COLOR research and application*, 11(Supplement):S15–S20, 1986.

[36] M. Livingstone and D. Hubel. Segregation of form, color, movement, and depth: Anatomy, physiology, and perception. *Science*, 240:740–749, May 6 1988.

[37] D. L. MacAdam. Uniform color scales. *Journal of the Optical Society of America*, 64:1691–1702, 1974.

[38] D. I. A. MacLeod. Computer controlled color displays in vision research: Possibilities and problems. *COLOR research and application*, 11(Supplement):S45–S46, 1986.

[39] W. T. Mayo. Using color to represent low spatial frequencies in speckle degraded images In B. E. Rogowitz, editor, *Proccedings of the SPIE, Volume 1077, Human Vision, Visual Processing and Digital Display*, 1077, 137–145, 1989.

[40] D. L. McShan and A. S. Glickman. Color displays for medical imaging. In H. J. Durrett, editor, *Color and the Computer*, chapter 10, pages 189–204. Academic Press, Orlando, FL, 1987.

[41] R. M. Merrfield. Visual parameters for color CRTS. In H. J. Durrett, editor, *Color and the Computer*, chapter 3, pages 63–82. Academic Press, Orlando, FL, 1987.

[42] G. W. Meyer and D. P. Greenberg. Color education and color synthesis in computer graphics. *COLOR research and application*, 11(Supplement):S39–S44, 1986.

[43] G. M. Murch. Color displays and color science. In H. J. Durrett, editor, *Color and the Computer*, chapter 1, pages 1–26. Academic Press, Orlando, FL, 1987.

[44] C. L. Novak and S. A. Shafer. Supervised Color Constancy for Machine Vision. In B. E. Rogowitz, M. H. Brill, and J. P. Allebach, editors, *Proceedings of the SPIE, Volume 1453, Human Vision, Visual Processing and Digital Display II*, 1453, 353–368, 1991.

[45] J. M. Olson. Color and the computer in cartography. In H. J. Durrett, editor, *Color and the Computer*, chapter 11, pages 205–220. Academic Press, Orlando, FL, 1987.

[46] C. A. Padgham and J. E. Saunders. *The Perception of Light and Colour*. Academic Press, New York, NY, 1975.

[47] R. M. Pickett and H. Levkowitz (Chairpersons). Panel: Visualization of multiparameter images. In A. Kaufman, editor, *Visualization '90*, pages 388–390, San Francisco, CA, October 23–26 1990. IEEE Computer Society, IEEE Computer Society Press.

[48] R. M. Pickett, H. Levkowitz, S. Seltzer, and C. D'Orsi. Integrated displays of multiparameter tomographic images. In J. T. Ferrucci and D. D. Stark, editors, *Liver Imaging: Current Trends in MRI, CT and US*, Boston, MA, June 25–27 1990. Massachusetts General Hospital and Harvard Medical School. Abstract.

[49] D. H. Pritchard. U.s. color television fundamentals—a review. *IEEE Transactions on Consumer Electronics*, CE-23(4):467–478, November 1977.

[50] J. M. Reising and A. J. Aretz. Color computer graphics in military cockpits. In H. J. Durrett, editor, *Color and the Computer*, chapter 8, pages 151–170. Academic Press, Orlando, FL, 1987.

[51] B. E. Rogowitz. The human visual system: A guide for the display technologist. *Proceedings of the SPIE* 24/3,235-252.

[52] B. E. Rogowitz. The psychophysics of spatial sampling. In G. H. Hugnes, P. E. Mantey, B. E. Rogowitz, editors, *Proceedings of the SPIE, Volume 901, Image Processing, Analysis, Measurement and Quality*, 901, 130–138, 1988.

[53] B. E. Rogowitz, D. T. Ling, and W. A. Kellogg. Task dependence, veridicality and pre-attentive vision: Taking advantage of perceptually-rich computer environments. In B. E. Rogowitz, editor, *Proceedings of the SPIE, Volume 1666, Human Vision, Visual Processing, and Digital Display, III*, 1666, in press.

[54] R. F. Sapita. Process control using color displays. In H. J. Durrett, editor, *Color and the Computer*, chapter 6, pages 115–138. Academic Press, Orlando, FL, 1987.

[55] C. Schmandt. Color text display in video media. In H. J. Durrett, editor, *Color and the Computer*, chapter 14, pages 255–266. Academic Press, Orlando, FL, 1987.

[56] J. J. Sheppard, Jr. Pseudo-color as a means of image enhancement. *American Journal of Optometry and Archives of American Academy of Optometry*, 46:735–754, 1969.

[57] L. D. Silverstein. Human factors for color display systems: concepts, methods, and research. In H. J. Durrett, editor, *Color and the Computer*, chapter 2, pages 27–62. Academic Press, Orlando, FL, 1987.

[58] W. Smith. Ergonomic vision. In H. J. Durrett, editor, *Color and the Computer*, chapter 5, pages 101–114. Academic Press, Orlando, FL, 1987.

[59] H. A. Spielman. Color and business graphics. In H. J. Durrett, editor, *Color and the Computer*, chapter 15, pages 267–284. Academic Press, Orlando, FL, 1987.

[60] M. C. Stone. Color, graphics design, and computer systems. *COLOR research and application*, 11(Supplement):S75–S82, 1986.

[61] D. J. Struik. *Differential Geometry*. Addisson-Wesley, Reading, MA, 1961.

[62] D. F. Switzer and N. C. Nanda. Doppler color flow mapping. *Ultrasound in Medicine and Biology*, 3:403–416, 1985.

[63] D. Travis. *Effective Color Displays*. Academic Press, 1990.

[64] M. W. Vannier and D. Rickman. Multispectral and color-aided displays. *Investigative Radiology*, 24(1):88–91, January 1989. Current Concepts, S. E. Seltzer, Ed.

[65] C. Ware and J. C. Beatty. Using color to display structures in multidimensional discrete data. *COLOR research and application*, 11(Supplement):S11–S14, 1986.

[66] K. L. Weiss, S. O. Stiving, E. E. Herderick, J. F. Cornhill, and D. W. Chakeres. Hybrid color MR imaging display. *American Journal of Roentgen Ray Society (AJR)*, 149:825–829, October 1987.

[67] M. E. Wigert-Johnston. Color graphic displays for network planning and design. In H. J. Durrett, editor, *Color and the Computer*, chapter 7, pages 139–150. Academic Press, Orlando, FL, 1987.

[68] W. D. Wright. A re-determination of the trichromatic coefficients of the spectral color. *Transaction of the Optical Society*, 30:141, 1928–29.

[69] W. D. Wright. *The Measurement of Colour.* Van Nostrand Reinhold Company, London, 1969.

[70] T. Young. On the theory of light and colours. *Philosophical Transaction of the Royal Society of London*, 92(1):12–48, 1802.

INDEX

1-19-09

32.90